Animal Signals

JOHN MAYNARD-SMITH

DAVID HARPER

School of Biological Sciences
University of Sussex

OXFORD
UNIVERSITY PRESS

Great Clarendon Street, Oxford OX2 6DP

Oxford University Press is a department of the University of Oxford.
It furthers the University's objective of excellence in research, scholarship,
and education by publishing worldwide in

Oxford New York

Auckland Bangkok Buenos Aires Cape Town Chennai
Dar es Salaam Delhi Hong Kong Istanbul Karachi Kolkata
Kuala Lumpur Madrid Melbourne Mexico City Mumbai Nairobi
São Paulo Shanghai Taipei Tokyo Toronto

Oxford is a registered trade mark of Oxford University Press
in the UK and in certain other countries

Published in the United States
by Oxford University Press Inc., New York

First edition published 2003

A catalogue record for this title is available from the British Library

Library of Congress Cataloguing in Publication Data
(Data available)

ISBN 0 19 852684 9 (Hbk)

ISBN 0 19 852685 7 (pbk)

10 9 8 7 6 5 4 3 2 1

Typeset by Newgen Imaging Systems (P) Ltd., Chennai, India
Printed in Great Britain
On acid free paper by Biddles Ltd, www.biddles.co.uk

Preface

Why are animal signals reliable? This is the central problem for an evolutionary biologist interested in signals. Of course, not all signals are reliable: but most are, otherwise, receivers of signals would ignore them. A number of theoretical answers have been proposed, and empirical studies made, but there still remains an immense amount of confusion. There are a number of reasons for this, but we think the major factor is that many biologists assume that there is only one correct explanation for reliability. We believe that there are a number of different answers, relevant in different cases, and that we must learn to distinguish between them. In this book, we attempt to explain the different theories, to give examples of signalling systems to which one or other theory applies, and to point to the many areas where further work, theoretical and empirical, is required.

One cause of confusion has been that different authors have used the same term with different meanings, and different terms with the same meaning. We have been guilty of this ourselves in the past. In this book we have tried to define our terms precisely, and to be consistent in their use. There is no way in which scientists can legislate about the meanings of theoretical terms. However, we have tried to reach a consensus. Before starting, we wrote to a number of workers in the field, outlining our intentions and asking for comments. We received helpful replies from Drs Carl Bergstrom, Marion Dawkins, Magnus Enquist, Alan Grafen, Tim Guilford, and Oren Hasson. We hope that the terminology we have adopted does not conflict with that favoured by any or these authors in a way that will cause confusion. We have kept in touch with them while writing the book and have recieved much helpful advice and information. We have also received helpful input from Drs Richard Andrew, Tecumseh Fitch, Kevin Fowler, Rufus Johnstone, Karen McComb, Andrew Pomiankowski, Anne Pusey, Francis Ratnieks, Vernon Reynolds, Tim Roper, Tom Seeley, and Joan Silk. We thank them all.

Contents

1 Introduction: what are signals? 1

1.1 Introduction 1
1.2 Signals and cues 3
1.3 The problem of reliability 6
1.4 The forms of signals 9
1.4.1 Efficacy 9
1.4.2 Evolutionary origin 9
1.4.3 Mimicry 10
1.4.4 Sensory manipulation 10
1.5 Action–response games, and some alternatives 11
1.6 Sexual selection 12
1.6.1 Paternal care 13
1.6.2 Sexually attractive sons 13
1.6.3 'Good genes'—indices and handicaps 14
1.6.4 The female gets nothing (sensory exploitation) 14
1.7 Definitions and terminology 15

2 The theory of costly signalling 16

2.1 Introduction 16
2.2 A brief history of the handicap principle 17
2.3 The Philip Sidney game 20
2.3.1 The discrete model 21
2.3.2 A model with continuously varying signallers 22
2.3.3 A model with continuously varying signals and responses 24
2.3.4 Perceptual error 25
2.3.5 Conclusions 26
2.4 'Pooling equilibria'—a more radical proposal 27
2.5 Non-signalling equilibria 29
2.6 Must honest signals always be costly? 29
2.7 Conclusions 30

3 Strategic signals and minimal-cost signals 32

3.1 Introduction 32
3.2 Strategic signals 33
3.2.1 Stalk-eyed flies 33
3.2.2 Musth in elephants 34
3.2.3 Chick begging 35
3.3 Minimal-cost signals 37
3.3.1 When can minimal-cost signals be evolutionarily stable? 37
3.3.2 Signals between unrelated individuals with a common interest 39
3.3.3 Relatedness 41
3.3.4 Kin recognition 42
3.4 Conclusions 44

4 Indices of quality 45

4.1 Introduction 45
4.2 Are mammalian sounds reliable indices of size? 45
4.3 The evolution of indices 47
4.4 Indices in different contexts 48
4.4.1 Indices of condition 48
4.4.2 Indices of size and RHP 50
4.4.3 Performance indices 51
4.4.4 Parasites 53
4.4.5 Indices of ownership 54
4.4.6 Signals in contests, and in mate choice 59
4.5 Indices and handicaps 60
4.6 Some problem cases 61
4.6.1 Stotting 61
4.6.2 Fluctuating asymmetry 63
4.6.3 Displays of weapons 66

5 The evolution of signal form 68

5.1 Ritualization 68
5.2 Efficacy 73
5.3 Arms races, manipulation and sensory bias 74
5.3.1 Introduction 74
5.3.2 A model, and an experiment 76
5.3.3 The response to novel signals 77
5.3.4 The comparative data 80
5.3.5 Conclusions 81

5.4 Sensory manipulation 81
5.4.1 Frogs and swordtails 81
5.4.2 Nuptial gifts in insects 84
5.4.3 Further examples of sensory manipulation 85
5.5 Mimicry and cheating 86

6 Signals during contests 90

6.1 Introduction 90
6.2 Badges of status 92
6.2.1 An avian example 92
6.2.2 ESS models of badges 94
6.2.3 Conclusions 95
6.3 Can signals of need settle contests? 96
6.3.1 The war of attrition 96
6.3.2 The war of attrition with random rewards 97
6.3.3 A model of conventional signals of need 97
6.3.4 Conclusions 98
6.4 Punishment 99
6.5 Protracted contests and varied signals 100
6.5.1 Varied signals—the evidence 101
6.5.2 Cichlid fishes and the sequential assessment game 102
6.5.3 Spider fights, and a motivational model 104
6.5.4 Territorial behaviour and the negotiation game 108
6.6 Conclusions 109

7 Signals in primates and other social animals 112

7.1 Introduction 112
7.2 Vervet Monkeys: a case study 113
7.3 How does the ability to signal develop? 116
7.4 Questions about what is going on in an animal's head 118
7.4.1 Do signals convey information about the external world? 118
7.4.2 Do signallers intend to alter the behaviour of receivers? 119
7.4.3 Conclusions 121
7.5 Social reputation and the honesty of signals 121
7.5.1 Introduction 121
7.5.2 A model 122
7.5.3 Evidence for direct reputation 123
7.6 Emotional commitment 124
7.6.1 Cultural and innate behaviour 124
7.6.2 'Altruistic punishment' in humans 125

7.6.3 Mutual displays 126
7.6.4 The interpretation of group displays 128
7.7 Human language 130
7.7.1 Cultural inheritance in Chimpanzees 131
7.7.2 The peculiarities of human language 132
7.7.3 The evolution of language 133

Glossary of scientific names 137

References 141

Index 161

1

Introduction: what are signals?

1.1 Introduction

This introduction can be treated as a menu outlining what is to come. There is no shortage of recent reviews of animal signals: for example, books by Hauser (1996), Bradbury and Vehrencamp (1998) and, more idiosyncratically, Zahavi and Zahavi (1997), and reviews such as Hasson (1994), Johnstone (1997), and Maynard Smith and Harper (1995). What need can there be for yet another? The answer is that despite, or perhaps because of, the multitude of reviews, there is widespread and often unrecognized confusion about the kinds of signal that exist, the mechanisms responsible for their evolution, and the terms to be used to describe them. The aim of this book is to clarify some of these confusions. We do not report any new facts, and in general, we rely on previously published theories; we do attempt to bring order out of chaos.

To illustrate how deep and often unconscious the confusions are, we start with an example. Enquist (1985) introduced the notion of a 'performance-based' signal, in which reliability is guaranteed because 'there is a relationship between the factor communicated and performance which is not easily removed'. As examples, he quotes the work of Davies and Halliday (1978) on the depth of croaks given by Common Toads, and of Clutton-Brock and Albon (1979) on the pitch of roaring in Red Deer (Fig. 1.1): in both cases, the quality of the sound is causally related to the size of the signaller, and hence to its fighting ability. In this book we use the word 'index' for such unfakeable signals: as another simple example, Thapar (1986) reports that Tigers mark their territories by scratching as high as they can on a tree trunk—an index of size that cannot be faked. We distinguish between indices and 'handicaps' (Zahavi 1975), which are signals whose reliability depends on the fact that they are costly to make. This distinction between index and handicap is the topic of the next two chapters, so we shall not explain it further here. Our immediate point is different. Enquist chose the calls of toads and of Red Deer as his examples of signals he referred to as 'performance-based', and which we are calling 'indices': in their recent book, Bradbury and Vehrencamp (1998) choose precisely the same two examples as illustrations of 'handicaps'.

Several explanations are possible for such an apparent confusion. Perhaps we are wrong in thinking that Enquist's notion of a performance-based signal is the same as our notion of an index. Or, perhaps, there really is no difference between the evolutionary mechanisms leading to these two kinds of signal. We do not think either

Fig. 1.1 Red Deer roaring, from Clutton-Brock *et al.* (1982).

of these arguments is tenable, but we may be mistaken. More plausibly, there may indeed be a distinction between the two mechanisms, and the disagreement is really an empirical one. For example, it may be that any stag could, at a cost, give any roar (but see p. 45), and the reason why there is a correlation between the pitch of roaring and fighting ability is that roaring is too costly for weaker stags. If so, the signal is indeed a handicap, and Enquist was wrong to use it as an illustration of a performance-based signal. So it may be that a disagreement about terminology in a particular case is not about theories, or the words used to describe them, but about what the world is like.

We have spent some time on this example, partly to illustrate the need for clarification in the study of animal signals, and partly, because it will help us to explain our own approach. We have suggested that the disagreements may be essentially theoretical—are handicaps and indices really different phenomena? Or they may be semantic—are we using the same words to mean different things? Or finally, they may be empirical—is a particular signal an example of a handicap, or an index, or neither? We think that the first requirement for clarity is the formulation of a model, usually, but not always in mathematical language. It is then possible to see whether the conclusions—for example, that a reliable signal must be costly—do in fact follow from the assumptions. Perhaps more important, given such a model it is possible to find out what is being assumed, even if the person formulating the model was unaware of it. For example we describe (p. 18) two models of sexual selection (Maynard Smith 1976; Pomiankowski 1987), both intended to investigate Zahavi's (1975) verbal model of the handicap principle, but leading to different conclusions. Given the models, it is fairly easy to spot the difference in the assumptions responsible for the different conclusions, but without formal models it would be hard to do so.

Given a formal model, it is possible to define terms such as handicap, index, cost, cue, and so on in a relatively unambiguous way. We think that terms are best defined in the context of specific models, rather than of examples. If one defines a term such as 'handicap' by saying 'by a handicap I mean a signal like the roaring of Red Deer', then if new empirical findings alter our view of the evolution of roaring, the meanings of our terms will change as well. In practice, of course, we have to think in terms of real examples as well as of abstract models. In this book we will illustrate models by concrete examples which we hope correspond: that is, that the assumptions of the model hold for the example. There is, therefore, always a risk that we are making assumptions about a particular example which will turn out to be false. But this is a risk biologists must always take if they want to relate their theories to the real world.

1.2 Signals and cues

We define a 'signal' as any act or structure which alters the behaviour of other organisms, which evolved because of that effect, and which is effective because the receiver's response has also evolved.

This definition has some implications that need spelling out. First, if a signal alters the behaviour of others it must, on average, pay the receiver of the signal to behave in a way favourable to the signaller; otherwise receivers would cease to respond. Thus, the definition distinguishes a signal from coercion. If one stag pushes another stag backwards, that is not a signal but coercion. If it roars and the other stag retreats, it is a signal, because the response depends on evolved properties of the brain and sense organs of the receiver. It follows that the signal must carry information—about the state or future actions of the signaller, or about the external world—that is of interest to the receiver. This information need not always be correct, but it must be correct often enough for the receiver to be selected to respond to it. Krebs and Dawkins (1984) presented animal signalling as an arms race between signallers as 'manipulators' and receivers as 'mind-readers'. When, as is usually the case, there is some conflict of interest between signaller and receiver, this picture must be correct. But it must also be true, that when natural selection has perfected both manipulation and mind-reading, both partners must on average benefit from the exchange; otherwise, the signalling system would cease to exist.

Second, the requirement that a signal evolved *because* of its effect on others distinguishes a signal from a 'cue', a term first used by Lorenz (1939). Hasson (1994) defines a cue as any feature of the world, animate or inanimate, that can be used by an animal as a guide to future action. The distinction between a cue and a signal is best illustrated by an example. Riechert (1978) studied contests between funnel-web spiders, *Agelenopsis aperta*, over web sites. She found that if there was a difference in weight between two spiders of 10% or more, the smaller spider retreated without risking a fight. A spider can perceive its weight relative to that of an opponent because the contests take place on the web. The spiders signal by vibrating the web, transmitting information about their size: a smaller spider can be converted into a winner by attaching a weight to its back. Thus, size itself is not a signal by our definition. It did not evolve *because* of its effect on other spiders, and in any case, most size differences between spiders are probably not genetic, but due to age or nutrition. However, the act of vibrating the web *is* a signal if, as seems plausible, it evolved because of its effect on the behaviour of an opponent through the information it provides about size.

Since, as often as not, a spider is smaller than its opponent, one might ask why it should risk informing its opponent of this fact? We discuss this question in Chapter 3: for the present, we want only to argue that size itself is not a signal, although it may be a cue, but that an act conveying information about size is a signal. Returning to the definition, the crucial point is that the signal must be able to evolve independently of any quality of the signaller about which it conveys information. It may help to give an example of a cue which is not a signal. A mosquito seeking a mammal to bite will fly up wind if it detects CO_2 (Gillies 1980). Thus, the CO_2 is a cue for the mosquito, but it is certainly not a signal by the mammal, which would prefer not to be bitten.

A few other examples will help to clarify the distinction between a signal and a cue:

1. *Courtship feeding.* This is, perhaps, a signal, because the habit of feeding a female during courtship can evolve independently of the ability to find food, and so may provide information about ability that can influence the behaviour of females, but courtship feeding may also evolve because it increases the fecundity of the female, and hence the fitness of the male; see p. 84.
2. *Fluctuating asymmetry (FA).* This is a term used for a random non-directional departure from perfect bilateral symmetry. The topic is discussed on pp. 63–5. There is evidence both for and against the view that a low degree of FA is associated with high fitness. If true, FA would be a good guide to female choice of a mate, particularly in a species with male parental care. However, even if FA was used in female choice, it would be a cue, not a signal. Only if some animal performed a display whose function was to make the level of FA more apparent would we speak of a signal.
3. *Aposematism.* A warningly coloured insect is sending a signal to a predator that it is distasteful. Note that the colouration can evolve independently of the quality, distastefulness, that is being signalled. Hence, the signal need not always be honest: a 'Batesian' mimic is an edible prey which resembles a distasteful and warningly coloured one (Fig. 1.2): it is signalling dishonestly. Depending on the relative costs and benefits to the predator of eating a distasteful prey, and consuming a palatable one, there will be a limit to the frequency of mimics relative to models before the warning colouration ceases to be effective in deterring predators.

Fig. 1.2 Batesian mimicry of various distasteful model species (on the left) by the edible polymorphic butterfly *Papilio memnon* (on the right). Five tightly linked genes determine, respectively, tails on the hindwing, white on the hindwing, white on the forewing, colour of 'shoulder patch', and colour of abdomen (from Turner 1984).

4. *Camouflage.* This is a doubtful case. As Hasson (1994) has argued, camouflage is a feature that alters the behaviour of another organism (a predator is less likely to eat a camouflaged prey), and that evolved for that reason; he, therefore, treats it as a signal. However, our definition of a signal requires that there be an evolved response, a requirement that is not met by camouflage. By response we mean a behavioural one. We do not deny that some predators have evolved counter-adaptations to camouflage. For example, sharks can use their electromagnetic receptors, the ampullae of Lorenzini, to detect cryptic or buried prey (Kalmijn 1982). But no predator could evolve a behavioural response to something it cannot detect, so camouflage does not meet our definition of a signal.

5. *Angler-fish lures.* This is certainly a signal by our definition. However, note that it is effective only because lures are rare compared to worms.

Recently, some authors have redefined the term 'cue'. The most drastic suggestion was that of Hauser (1996): both cues and signals are traits that have evolved because they alter the behaviour of other individuals, but they differ in two ways. First, cues are permanently 'on', while signals are switched 'on' and 'off' depending on the circumstances. Second, once a cue has been produced it costs nothing extra to express it, whereas signalling can impose additional costs. As examples of cues, Hauser (1996) used warning colouration and 'badges of status'—patches of colour that correlate with dominance (Section 6.2). Completely redefining a term such as cue with a well-established usage (Lorenz 1939) seems to us to be a certain route to confusion. Moreover, the distinction that Hauser (1996) made is far from clear cut. Permanent colour patterns can often be concealed. For example, many Arctiid moths, which are defended by pyrrolizidine alkaloids, hide their brightly coloured hindwings and abdomen beneath cryptically coloured forewings until disturbed by a potential predator (Cardoso 1997). Indeed, Hauser (1996) emphasized the distinction (originally made by Hansen and Rohwer 1986) between 'fixed badges' of status which he regarded as cues and 'coverable badges' which he regarded as signals. But it is untrue that in the first case 'individuals cannot control the degree to which they express or show off their badge'. Great Tits show off their belly stripe, which acts as a badge of status (e.g. Jarvi and Bakken 1984), in elaborate displays (e.g. Wilson 1992*a*). Male House Sparrows even fluff up their badge (a black bib under the bill, Møller 1987*a*), increasing its apparent size (Cramp and Perrins 1994). We agree that there are interesting differences between a signal such as a warning colouration that lasts a lifetime and one like a call that lasts a fleeting second. But these are the ends of a continuum of signal duration, and not a dichotomy.

A less radical redefinition of 'cue' is that of Galef and Giraldeau (2001). They agreed that cues have not evolved to alter the behaviour of other animals, but restricted the term to behavioural traits. Their example was of the sound of one animal eating attracting other individuals. They distinguished 'cues' such as this from 'signs' which are residual consequences of behaviour which alter other individuals' behaviour. For example, food remains, faeces, and footprints might all attract animals to a foraging site. But, as with Hauser's more radical redefinition, this adds semantic confusion to a field in which it is rife, and we are reluctant to accept it.

1.3 The problem of reliability

We now turn to some of the main problems discussed. First, what maintains the reliability, or 'honesty', of signals? This question has been central to theories of the evolution of signals since Zahavi (1975) proposed the 'handicap principle'. It is intriguing that the problem of honesty is not peculiar to communication between animals. In his book on semiotics, Eco (1976) wrote 'semiotics is in principle the discipline studying everything that can be used in order to lie'. An account of the handicap principle, and of alternative reasons why signals may be reliable, which reaches conclusions very similar to those proposed in this book, is given by LaPorte (2002).

The examples of mimicry and aposematism make it clear that a given signal does not have to be honest always. But if a signal is to be effective in eliciting the appropriate response, it must be honest most of the time, and it is this that has to be explained. Zahavi's proposal was that signals are honest because they are costly to make (Fig. 1.3). Before discussing this idea further, it is important to emphasize that not all signals are costly to make (Fig. 1.3), or rather, they are no more costly than is necessary to convey the necessary information. Thus, it is necessary to distinguish two kinds of 'cost'—'efficacy cost' and 'strategic cost' (Guilford and Dawkins 1991). Efficacy cost is that needed to transmit the information unambiguously, and must be paid even when the signaller has no temptation to lie. Following Grafen (1990*a*), we refer to the additional cost needed to maintain the honesty of a signalling system as 'strategic cost'.

Strategic costs can arise in several ways. First, the signal may be a structure whose development is costly in resources: for example, in Section 3.2.1 we argue that the long eye stalks of male stalk-eyed flies are costly for this reason, and should be interpreted as handicaps. The tail of a peacock may be costly for this reason: it may also be costly in exposing its owner to greater risks of predation. Some signals made during contests may make the signaller more vulnerable to attacks by its opponent. For example, Enquist *et al.* (1985) investigated the significance of risky signals between Northern Fulmars in contests over food. The contests were between the 'owner' of a

Fig. 1.3 The contrast between 'notices' and 'advertisements', from Maynard Smith (1958), who argued that notices are no more expensive than is needed to convey the necessary information (i.e. efficacy cost), whereas advertisements are costly because they must compete with rival signals: today, the latter would probably be interpreted as 'handicaps'.

Fig. 1.4 The breast-to-breast display, made during contests between Fulmars over pieces of fish (from Enquist *et al.* 1985): an example of a 'risky' signal.

resource—a piece of fish—and a rival. The birds made a range of different signals, some of which (e.g. the breast-to-breast display; Fig. 1.4) risked an escalated contest. The authors found that the more risky signals were more effective in causing an opponent to withdraw. They argue that individual differences responsible for the choice of signal concern need for the resource rather than fighting ability. Strategic costs may be of any of these kinds, and the corresponding signals are best regarded as handicaps.

There is, however, one context in which we prefer not to call a signal a handicap, although it may have costly consequences. Suppose there is 'punishment' (see Section 6.4). An individual indicates that it will do A, but then does B, and is punished by the receiver: lying signals are costly. Our reason for not referring to such cases as handicaps is that at an evolutionary equilibrium, lying will be rare or absent, and signals can be both cost-free and reliable. The problem lies, not in explaining reliability, given punishment, but in explaining why the receiver is willing to incur the cost of punishing.

Chapter 2 is devoted to an analysis of the handicap principle. Following Pomiankowski (1987) and Grafen (1990*a*), it shows that there are situations in which signals have to be costly in order to be honest. However, there are at least two other mechanisms that can give rise to a stable and reliable signalling system. The first is that signaller and receiver have a common interest in the outcome of the interaction. For example, a warningly coloured toxic insect and a potential predator both benefit if the insect is not attacked. In Chapter 3 we discuss examples of signalling systems that are reliable because of the strategic cost of signalling, and others that are reliable because of common interest.

The second alternative to the handicap principle as a mechanism ensuring reliability has already been mentioned, for example, in the case of spiders signalling their weight. The signal may be an index of some quality, and is honest because the signaller cannot lie. This explanation for honesty was discussed by Maynard Smith and Parker (1976), using the term 'assessment signals'; here, we prefer to use the term 'index' because both indices and handicaps are signals that make assessment easier. Possible examples of indices, and the difficulty of distinguishing them from handicaps, are discussed in Chapter 4.

Thus, there seem to be three alternative explanations for the reliability of signalling systems—the handicap principle, common interest, and indices of quality. Are there any others? We think that there is certainly one other, and perhaps two. The one we are reasonably confident of we will call 'reputation'. In a species, and ours may not be the only one, with sufficient powers of individual recognition and memory, lying may not pay because it will be remembered against you. Reputation is relevant in primates, and to some degree in other groups. The special properties of primate signalling systems are discussed in Chapter 7.

A final possibility may not occur in animals, but should be mentioned. Classical game theorists have pointed out that it may benefit a player to bind himself beforehand to a particular course of action. Ulysses wisely bound himself to the mast before sailing past the sirens. It would pay a Union leader to get his members to vote for '10% or we strike' before negotiations start, provided he was sure the employer would rather pay 10% than suffer a strike. The logic is clear, but can animals follow it? What is needed is a signal which, honestly, binds an animal to a particular course of action. Maynard Smith (1982) suggested that musth in elephants might be an example of this mechanism, but, as explained on p. 34, this is probably wrong.

1.4 The forms of signals

Chapter 5 discusses why signals have the particular forms that they do. A number of factors are relevant.

1.4.1 Efficacy

Like other features of organisms, signals are subject to design requirements.

1.4.2 Evolutionary origin

A major problem in the evolution of signals is that, even if a model predicts a stable signalling equilibrium, a 'no-signalling' equilibrium is usually a stable alternative: that is, if no one signals, do not evolve the capacity to respond, and if no one responds, do not bother to signal. So how does signalling get started?

Tinbergen (1952) suggested that signals originate as ritualized intention movements: to raise one's fist is perceived as aggressive because it is an exaggerated version of the initial stage of hitting someone. In mammals, the baring of teeth, or lowering of horns, may occur as signals that are not necessarily followed by actual attack. Bird chicks often beg for food by opening their beaks, revealing their brightly coloured mouths. In such cases, the form of the signal has a direct relation to its meaning. However, signals may also originate by the ritualization of movements that have no obvious relation to their present meaning. For example, in ducks some courtship movements made by courting drakes appear to be exaggerations of preening movements (see p. 71). In such cases, the form of the signal has no obvious relation to its meaning: it is a symbol rather than an icon (see p. 11). Probably such signals originate as exaggerations of 'displacement activities' (see p. 71).

The term 'ritualization' is also used to refer to the kinds of change that occur when a cue is converted into a signal. Signals differ from cues in four main ways (Wiley 1983; Johnstone 1997): they tend to be more conspicuous, redundant and stereotyped, and they are often preceded by alerting components. The conversion of cues into signals, and the kinds of change that occur during the conversion, are discussed further in Chapter 5.

1.4.3 Mimicry

It is customary to divide mimicry into two categories—Mullerian and Batesian. In the former, two or more distasteful species have similar warning colouration. In both species the signal 'I am distasteful' is honest. Both species benefit, because a potential predator has more opportunities to learn the meaning of the signal. In contrast, a Batesian mimic is palatable, but mimics the colouration of a distasteful model species: thus, it gives an unreliable signal. There are many examples of an organism dishonestly mimicking an honest signal given by another species. Early Spider Orchids attract pollinators by producing 14 out of the 15 chemicals in a pheromone produced by female bees, *Andrena nigroanea* (Schiestl *et al.* 1999). Bolas spiders (*Mastophora* sp.) emit a pheromone which lures army worm moths *Spodoptera frugiperda* to within capture range (Eberhard 1977). In some cases, the model for a dishonest signal is a cue, not a signal given by another species. For example, an angler fish attracts prey by means of a lure resembling a worm, and the salticid spider *Portia fimbriata*, which preys on orb-web spiders, attracts its victim by vibrating its web in a manner that mimics the struggles of a captured fly (Tarsitano *et al.* 2000).

1.4.4 Sensory manipulation

In cases of mimicry, the unreliable signal is believed because it resembles a reliable cue or signal. Its effectiveness depends on being rare compared to the model. Are there signals that are unreliable but effective, yet do not owe their effectiveness to their resemblance to a reliable cue or signal? Krebs and Dawkins (1984) suggested that signalling could best be analysed as an arms race between signallers attempting to manipulate receivers to their own advantage, and receivers attempting to mind-read the signallers. There has to be some truth in this picture. Some manipulation takes the form of mimicry of the kind just described, as in the case of brood-parasitic indigo birds (*Vidua* spp.) mimicking the songs and nestling gape patterns of the estrildid finches which foster them (Payne and Payne, 1994). But it has recently been argued

(Ryan 1990) that in some cases manipulation depends on a pre-existing sensory bias in the receiver which may be accidental, rather than a response that is adaptive in other circumstances. To demonstrate this, it is necessary to study related species, with a reliable phylogeny (see Section 5.4).

The theory of sensory manipulation provides an explanation of signalling that is in sharp contrast to the optimization approach offered by game theory, which assumes that an evolutionary equilibrium has been reached in which each 'player', signaller and receiver, does the best it can, given what the other is doing. In contrast, according to the manipulation theory the sensory system is such that, even if a temporary equilibrium is reached, there will often be some new signal that evokes a more favourable response than the one being used. Several lines of evidence have been quoted in support of this view. The ethologists pointed to 'supernormal stimuli', rarely or never experienced, that are more effective than the normal one. For example, oystercatchers retrieve eggs lying outside the nest by rolling them under their chin: the adults are much more likely to retrieve a model egg if it is unusually large, even three times normal volume (Tinbergen 1948). Experimental psychologists have found that if an animal is trained to respond to stimulus X, it will sometimes respond more strongly to a stimulus slightly different from X, which, if repeated, may lead to the evolution more of extreme signals (Weary *et al.* 1993).

In discussing the forms of signals, three terms used in semiotics—index, icon, and symbol—can be useful. An index is a signal whose form and intensity are physically associated with some quality of interest to the receiver: in effect, it is an unfakeable signal. We have already discussed indices in animal communication; the term 'performance-based signal' (Enquist 1985) also refers to signals that cannot be faked. An icon is a signal whose form is similar to its meaning. Familiar icons in human communication are those that appear on computer screens, or on the doors of public lavatories. Among animals, the classic example is the dance of honey bees, in which the orientation of the dance indicates the direction and its length the distance of a food source. However, in so far as many animal signals are ritualized intention movements, they are likely to have an iconic character. Finally, a symbol is a signal whose form is arbitrarily connected to its meaning. Most human words are symbolic. It is arbitrary that a 'window' is an opening to look through and 'door' one to walk through—it could be the other way round. Many words probably started out as icons, and a few have retained that character, for example, many bird names such as Peewit, Curlew, Killdeer, Chiffchaff; happily, even some binomial names have retained this onomatopoeic character, for example, *Crex crex* and *Upupa epops*. The term 'conventional' has been used in many different senses by writers on animal signals (Guilford and Dawkins 1995): in this book it is used with the same meaning as symbolic.

1.5 Action–response games, and some alternatives

The theory of signalling is often discussed in the context of a simple 'Action–Response Game' between two individuals, one of which sends a signal to which the other responds. The interaction is treated as a 'game' because, as evolutionary biologists, we expect each partner to the interaction—signaller and receiver—to perform an action which is optimal, given what the other partner can be expected to do. We have already mentioned one context—sensory manipulation—in which such an approach is inappropriate. There are three other contexts in which the simple action–response model runs into difficulties: contest behaviour, social behaviour, and sexual selection.

Chapter 6 discusses signals made during contests over territory, food, mates, or position in a hierarchy. Here, the inadequacies of a simple action–response model are only too obvious. Such contests often consist of a long and varied sequence of signals by both participants. Long sequences are hard to model mathematically, so that we are forced to rely on crude and oversimplified models, and on intuition backed by rather hand-waving verbal arguments. From a theoretical point of view such protracted contests are a challenge, but some progress can be reported.

An action–response game may be an inadequate model of the behaviour of social animals, because if two individuals interact repeatedly, an animal's response to a signal may depend on its memory of how the signaller has behaved in previous interactions—an effect we refer to as 'reputation'.

A third context in which simple action–response games may be inadequate is that of sexual selection. It is true that courtship often consists of a set of signals by the male, to which the female responds either by mating, or not. However, difficulties arise in understanding the selective forces involved: these are discussed in the next section.

1.6 Sexual selection

Many of the signals described in this book are made during courtship. They occur in a number of different contexts, and show a corresponding range of forms. In this section we review these contexts, and refer to more detailed discussions later in the book.

Courtship is often asymmetrical—the male attempting to persuade the female to mate, and the female responding. This asymmetry arises because eggs are more expensive than sperm. In many species the male contributes nothing but sperm, whereas the female contributes eggs, and often cares for the young. In such cases, the male maximizes his fitness by mating as often as possible, and the female need mate only once, but may benefit by choosing the right male. However, this pattern is not universal: the male may contribute to the care of the young. For example, in many birds the two parents contribute equally. In a few cases, the male contributes more than the female. For example, in pipefishes and sea horses (family Syngnathidae) only males brood embryos in pouches that supply oxygen and nutrients. This paternal care is associated with sex-role reversal in some but—interestingly—not all species (Vincent *et al.* 1994; Jones *et al.* 2001). In pipefish, males limit female reproductive rate: as a consequence, females compete fiercely for mates, and males prefer ornamented females as partners (Berglund and Rosenqvist 2001).

The central problem in understanding sexual selection and the signals exchanged lies in deciding what the female is getting out of the interaction. 'Choice' does not contribute only to the female's survival—indeed, it may be costly—but to the survival of her offspring, and perhaps to their genotype, and hence, to the chance that she will have grandchildren. There is not a single answer to the question, 'what is the female getting out of it?', but several. However, the answer to the question is likely to influence the form of the signals employed. In this section, we review briefly the possible answers and refer to more detailed discussions of the possible answers later in the book.

1.6.1 Paternal care

The extent of paternal care varies enormously. In species in which the two parents contribute more or less equally, one might expect courtship also to be symmetrical. There are mutual displays (see Section 7.6.3), but these are usually concerned with preserving a pair bond already formed, or in ensuring cooperation, rather than in initial mate choice. In passerine birds, males usually contribute fully to feeding the young, but are often more brightly coloured than females, and song, which is concerned with establishing a territory and attracting a mate, is usually confined to males. One reason for this continued difference between male and female signals, even in species that form a pair bond that lasts through the breeding season and which share the costs of raising young equally, is that the asymmetry between the costs of eggs and sperm remains. Consequently, males have more to gain by mating with females other than their mates: 'extra-pair copulations' are, in fact, common in many species.

1.6.2 Sexually attractive sons

Fisher (1930) pointed out that if, in any species, most females prefer to mate with a male with some specific trait—for example, a long tail—it will pay an individual female to conform to this trend because, in this way, she can ensure that her sons have long tails, and hence, are likely to mate often. He suggested that this could lead to a 'runaway process'—that is, a continuing elaboration of the preferred trait, and of preference for it. This provides a possible explanation for the extreme elaboration of male secondary sexual characters in species in which males contribute nothing but sperm, particularly in lekking species.

Fisher's process requires an initial female preference. This raises no particular difficulty—for example, it could arise from some non-adaptive sensory bias (see later), or from selection for reproductive isolation during a speciation event. A more serious problem (Pomiankowski and Iwasa 1998) is that choice may be costly for the female: if so, the process will not lead to extreme exaggeration of the male trait, but to an equilibrium between female costs and benefits. However, whether one accepts Fisher's runaway argument or not, in most species the kind of selection he described is likely to act so as to reinforce male sexual signals, and female preferences.

1.6.3 'Good genes'—indices and handicaps

If there is additive genetic variance for fitness (in non-mating contexts) in the population, it will pay a female to mate with a fitter male, because her offspring will be fitter, and she will, therefore, transfer her genes to more grandchildren. There are two quite distinct mechanisms whereby she might make such a choice—'indices' and 'handicaps' (or performance-based and choice-based signals), described earlier.

In this context, an 'index' is a signal causally related to fitness. No signal can be an accurate measure of total fitness. But a female may choose an index of some component of fitness. On p. 51 we describe how female *Drosophila subobscura* select males according to their ability to 'dance' (Maynard Smith 1956)—a measure of neuromuscular coordination. Hamilton and Zuk (1982, see p. 53) suggest that bare patches of skin (wattles and combs) may may act as indices of parasite resistance. Courtship feeding may sometimes act as an index of a male's foraging ability (although other mechanisms may be responsible, see p. 52).

An alternative mechanism (Zahavi 1975) is that the male's signal may be a 'handicap', and too costly for males of low quality to produce. This possibility is discussed in Section 3.2.1, using stalk-eyed flies as an example (David *et al.* 2000).

In extreme cases, it is rather easy to decide whether a male signal is an index or a handicap. If it is an index, it should be relatively cost-free, and should correlate with some trait that contributes to fitness in contexts other than mating. If it is a handicap, it should be costly to produce—that is, it should reduce fitness in contexts other than mating. The dance of a male *D. subobscura* lasts only a few seconds, and is virtually cost-free: the vision and neuromuscular coordination required contribute to fitness in other contexts (they are similar in males and females). In contrast, the long eye stalks of male stalk-eyed flies are costly in contexts other than mating, and are absent in females. But there are difficulties that are discussed in Section 4.5.

1.6.4 The female gets nothing (sensory exploitation)

Females may have sensory preferences that are non-adaptive, or adaptive in contexts other than mating. If so, they may be exploited by courting males (Basolo 1990; Ryan 1990). This possibility is discussed in Section 5.4. There are some clear examples, but they are few, perhaps because, in order to demonstrate sensory exploitation, it is necessary to show, by phylogenetic analysis of a group of related species, that the sensory bias predated the male signal.

Writers on sexual selection have a curious tendency to interpret all cases as examples of the same evolutionary mechanism. There is no reason why this should be so. Male signals and female responses may have evolved by different selective mechanisms in different species, depending on different contributions by the two sexes in raising their offspring, and on the chances of evolutionary history. More than one of the mechanisms listed above may have played a role in a single species—for example, a signal that initially exploits a sensory bias in females may be exaggerated by the Fisherian process, or a signal initially acting as an index of some component of fitness may be exaggerated until it becomes a handicap. More than one of the selective processes described may also act simultaneously in a single species: for example, the process envisaged by Fisher will usually operate, even if it is not the determining factor.

1.7 Definitions and terminology

A major difficulty in writing this book has been that, even when different authors agree on the distinctions that need to be made between different categories of signal, they may use quite different terms to describe them. In the following table we define as clearly as we can the distinctions we intend to make between types of signal, and the terms we shall use.

Signal. An act or structure that alters the behaviour of another organism, which evolved because of that effect, and which is effective because the receiver's response has also evolved.

Cue (Hasson 1994). A feature of the world, animate or inanimate, that can be used by an animal as a guide to future action.

Ritualization (Tinbergen 1952). Evolutionary process whereby a cue may be converted into a signal.

Handicap (Zahavi 1975) (= strategic signal (Grafen 1990*a,b*). A signal whose reliability is ensured because its cost is greater than required by efficacy requirements; the signal may be costly to produce, or have costly consequences (e.g. vulnerability cost (Adams and Mesterton-Gibbons 1995)).

Cost. Loss of fitness resulting from making a signal. Includes:

(1) *Efficacy cost* (Guilford and Dawkins 1991). Cost needed to ensure that the information can be reliably perceived.

(2) *Strategic cost* (Grafen 1990*a,b*). Cost needed, by the handicap principle (Zahavi 1975), to ensure honesty.

Index (Maynard Smith and Harper 1995). A signal whose intensity is causally related to the quality being signalled, and which cannot be faked (= performance-based signal (Enquist 1985)).

Minimal-cost signal. A signal whose reliability does not depend on its cost (i.e. not a handicap), and which can be made by most members of a population (i.e. not an index).

Icon. A signal whose form is similar to its meaning.

Symbol (= conventional signal). A signal whose form is unrelated to its meaning.

2

The theory of costly signalling

2.1 Introduction

In 1975, Zahavi proposed that animal signals are reliable because they are costly, an idea he called the 'handicap principle'. Although the idea has since been extended (Zahavi and Zahavi 1997) to lengths that seem to us excessive, the basic idea has been justly influential. Our aim in this chapter is to clarify the concept, and the situations to which it can properly be applied.

First, the notion of a 'cost' needs discussion. The crucial distinction is between 'efficacy cost' (Guilford and Dawkins 1991), the cost necessary to ensure that the information can be reliably perceived, and 'strategic cost', the cost needed to ensure honesty and prevent cheating. The efficacy cost is easy to understand: a road sign on a motorway must be larger and more expensive than one on a side road because it must be read at a greater distance, and not because, otherwise, someone would be tempted to lie. Demonstrating that a signal is costly, therefore, is not evidence that it is a handicap. For example, male nightingales sing loudly for most of the night and lose 5–10% of their body mass while doing so (Thomas 2002). However, this heavy investment in song may be needed if males are to be heard by females travelling hundreds of metres above them. Rather it is the great virtuosity of a nightingale's song (Hughes *et al.* 2002) that would make us wonder about strategic costs. Similarly, in Zebra Finches the calls of both sexes, and the songs of males, are louder in the presence of increased levels of white noise (Cynx *et al.* 1998). Thus, the loudness of their vocalizations is determined by the need to be audible in varying circumstances. In many cases, it is hard to tell whether signallers bear strategic costs. Consider, for example, the observation that *Phylloscopus* warbler species inhabiting shady habitats are more brightly coloured than those from open areas, even when phylogeny is controlled for (Marchetti 1993). This can most readily by explained by higher efficacy costs in dim light (Johnstone 1997). It is also consistent, however, with a greater risk of predation in open habitats constraining plumage brightness. Within a warbler species, bright colours might have strategic costs, so that only the fitter individuals can afford to be brightly coloured; if this is the reason for dull plumage in open habitats, it is an example of a handicap. The need for extra work to test whether a signal involves strategic costs or not, as in this case, is no reason for failing to make the distinction. Efficacy costs have implications for the form of animal signals which are discussed in Chapter 4.

The notion of strategic cost is wholly different: it is the cost needed to prevent cheating. Consider, for example, a signal made during a contest over an item of food, indicating 'I am seriously in need of this food, and willing to fight for it if necessary'. If an escalated fight is expensive, such a signal, if believed, would often persuade an opponent to retreat, and thus ensure access to the food. There would, therefore, be something to be gained by making the signal, even if untrue. The handicap principle asserts that such cheating signals do not occur, or are made only rarely, because they are too costly to make, unless the signaller is truly in need. We describe formal models of such an interaction later. But two points are at once obvious:

1. For the suggested mechanism to be relevant, it must be *possible* for an individual not in need to make the signal. There are cases in which this is not so. For example, when Bald Eagles squabble over food, hungry birds display their empty crop at rivals, directly demonstrating the reason for their high motivation (Hansen 1986). It would be seriously misleading to refer to such a signal as a 'handicap' (although one of us has been guilty of doing so, in the phrase 'revealing handicap' discussed below). We refer to such signals as indices. A handicap, in contrast, is a signal it is possible, but unprofitable, to make dishonestly. More precisely, there must be genetic variance for the nature of the signal given, independent of the quality about which information is being conveyed.
2. The cost of a handicap must be greater than just the efficacy cost required to transmit the information unambiguously: we refer to such additional costs as 'strategic costs'.

Strategic costs can arise in different ways. There may be costs associated with producing the signal: for example, the building of a bowerbird's bower consumes time and energy that could be used in other ways. Alternatively, the consequences of making a signal may be costly; for example, in aggressive encounters a signal requiring close approach to an opponent is more risky than one made at a distance. Some signals may be costly for both reasons; a peacock's tail is costly to develop, and makes the bird more vulnerable to predators. However, all these could in principle act as strategic costs, although it is an empirical question whether in fact they do so.

We first describe, with a minimum of mathematics, the debates stimulated by Zahavi's 1975 paper, and some of the formal models suggested to clarify his ideas. We then present a particular model, the Philip Sidney game, which illustrates two points: first, an evolutionarily stable signalling system maintained by strategic costs is a logical possibility, and second, there are circumstances in which cost-free signals can be stable. Finally, we discuss possible examples of costly and cost-free signals.

2.2 A brief history of the handicap principle

Early responses to Zahavi's handicap principle were unfavourable, in part perhaps because they were set in the context of genetic models of sexual selection and female choice, in which costs and benefits are hard to evaluate, instead of game-theoretic models. Maynard Smith (1976) considered a genetic model of sexual selection with three loci, affecting respectively female choice, the presence or absence of a male ornament, and male viability. He concluded that the process did not produce the results claimed for it; that is, more viable males were not more likely to have the ornament. In retrospect, one can see that the reason for this negative conclusion lay in the assumption that the effects of the male ornament and viability genes combined multiplicatively: that is, the presence of an ornament lowered the fitness of high and low viability males in the same proportion.

Hamilton and Zuk (1982) argued that an effective method whereby females can choose a male of high fitness is by selecting a male free of parasites. They suggested that some male ornaments make it possible for females to detect the presence of parasites. In discussing their idea, they quote Zahavi (1975), and explicitly reject the use of the term handicap for the process they are describing, on the grounds that the female is seeking, not a handicap, but a signal of good health that 'cannot be bluffed'. Thus, they drew a clear distinction between what we are calling here, a handicap and an index. Unhappily, Maynard Smith (1985) referred to the Hamilton–Zuk process as a 'revealing handicap', a phrase that has not helped to clarify ideas.

An alternative interpretation of 'handicap', still in the context of sexual selection, was made by West-Eberhard (1979), and in a more formal model by Anderson (1986). The basic idea is that only males of high viability produce the ornament or signal; males of low viability do not pay the cost of producing the ornament, and do not get the benefit of female preference. This mechanism (referred to by Maynard Smith 1985, as a 'conditional handicap') seems to be closer to the spirit of Zahavi's original idea. However, a full understanding of the handicap principle, and of the role played by costs in maintaining a signalling equilibrium, arose from three papers, by Enquist, Pomiankowski, and Grafen, which we now describe.

Enquist (1985) started by drawing a distinction between the 'choice' of a signal and its 'performance'. By performance he meant what we are calling an index: there is a causal connection between the level of performance and the quality being signalled. He did not doubt that some signals are indices: as likely examples, he quoted the fact that Common Toads signal their size by the depth of their croaks (Davies and Halliday 1978) and that the roaring of Red Deer stags is an indication of their size and condition (Clutton-Brock and Albon 1979). However, he was more interested in 'choice', arguing that there are many cases in which an animal is capable of performing any one of a number of actions, and yet its choice of action influences the behaviour of others. Stimulated by Zahavi (1975, 1977), he sought for an explanation in terms of costs that may arise from making a particular signal.

He analyses a simple model of a contest between two animals over a valuable resource. Individuals differ in their fighting ability, and are supposed to know their own ability but not that of their opponent. An animal has a choice of two signals, A or B; these signals cost nothing to produce, although they may have costly consequences. He shows that there can be a signalling equilibrium, in which males of high fighting ability signal A, and attack if their opponent does not retreat, and males of low fighting ability signal B, and retreat at once, without injury, if their opponent signals A.

However, equilibrium requires that the strategy 'signal A, but retreat at once before you are injured if your opponent signals A' is not a possible option. This proviso is crucial, and needs some justification. It could be true because signal A involved a close approach to the opponent so that escape was difficult. It would also be true if an individual giving signal A, and then attempting to retreat if its opponent does likewise, was attacked. A role for the 'punishment' of individuals giving dishonestly aggressive signals has been suggested in other contexts (e.g. in 'badges of status', see Section 6.2). In Enquist's model, stability requires that the cost to a weak individual of signalling A—that is, of cheating—must actually be greater than the cost of an escalated fight between equals. In other words signal A is intrinsically risky, and costly for that reason. Indeed, for an equilibrium to exist, it must be more costly for a weak individual, a point whose relevance will become clearer below.

Enquist *et al.* (1985) investigated the significance of risky signals between Northern Fulmars in contests over food. These contests were between the 'owner' of a resource—a piece of fish—and a rival. The birds made a range of different signals, some of which (e.g. the breast-to-breast display; figure 1.4) risked an escalated contest. They found that the more risky signals were more effective in causing an opponent to withdraw. They argue that the individual differences responsible for choice of signal concern need for the resource, rather than fighting ability. This work is important because it distinguishes clearly between choice and performance (or handicap and index), it provides a clear model showing that a signalling equilibrium may require costly signals, and indicates the kind of field observations needed to demonstrate the handicap principle.

The handicap principle was investigated by Pomiankowski (1987) in the context of sexual selection. He investigated a genetic model with three loci, affecting respectively female preference, the presence of a costly male ornament, and male viability; thus, his model closely resembles that analysed with negative results by Maynard Smith (1976). He observed a runaway process of sexual selection leading to the fixation of the costly ornament, provided that either

(1) the fitness cost of the ornament was greater in males of low viability; this is the crucial difference from Maynard Smith's model; or
(2) the ornament is not costly to produce, but its presence makes it possible for the female to perceive the presence of the high viability allele.

Thus, he found a signalling equilibrium if either the signal was a handicap (case 1), or an index (case 2); in the case of a handicap, the cost of the signal had to be greater for individuals of low quality—a condition similar to that found by Enquist (1985).

The final step leading to a general acceptance of the handicap principle was the publication of two papers by Grafen (1990*a*,*b*). Like Pomiankowski, he analysed the concept in the context of sexual selection, but reached conclusions that hold more generally. He assumes that males vary in some quality, q, of interest to females. It pays females to mate with males of high quality, but they cannot perceive q directly. Males, however, give an 'advertisement', a, which can be perceived, and which is costly to produce. The phenotype of a male is specified by his quality, q, and by $a = A(q)$, his level of advertisement; thus $A(q)$ is a function by which a male converts his quality into advertisement. A male cannot alter his quality, q, but can alter a: that is, natural selection can alter his allocation of resources to advertisement for a given quality. The cost of advertisement depends on a and on q. The female has a rule for estimating the quality of a male from his advertisement, $p = P(a)$, where p is the perceived quality. Her fitness is a maximum when $p = q$, and falls off as the discrepancy between true and perceived quality increases.

The evolutionary variables, then, are the functions $A(q)$ in males, and $P(a)$ in females. An evolutionary equilibrium is a pair of functions, $A^*(q)$ and $P^*(a)$, such that, if males advertise according to A^*, then the best a female can do is to adopt the rule P^*, and, if females do adopt this rule, the best a male can do is to advertise according to the rule A^*.

For very general assumptions about the form of the functions determining the cost of a given advertisement and the price paid by a female for misinterpreting a signal, Grafen showed that, if a signalling equilibrium exists, it will have the following properties:

(1) a female can correctly infer a male's quality from his advertisement (honesty);
(2) signals are costly; and
(3) a given signal is more costly for a male of low quality (note that both Enquist and Pomiankowski found that this condition was necessary in particular cases).

This result is both general and important. However, it is capable of misrepresentation. Grafen (1990*a*) follows this list of conditions with the statement 'if we see a characteristic which does signal quality, then it must be a handicap'. What of a spider which vibrates a web, thereby conveying information about its weight to an opponent? It seems that Grafen would not regard this as a signal. He refers explicitly to 'the revealing handicap, which is not a handicap at all, nor a signal'. We agree that cost-free actions which convey accurate information about the actor are not handicaps, but we do regard them as signals. However, although it is easy to distinguish between a handicap and an index in the context of a model, it may be very difficult to do in real cases. We shall meet this difficulty repeatedly, particularly in the next chapter. For the present, we want only to insist that both handicaps and indices (or 'choice' and 'performance' in Enquist's terminology) are logical possibilities: whether we regard the latter as signals is ultimately a matter of semantic taste.

2.3 The Philip Sidney game

This game was originally suggested (Maynard Smith 1991) to provide a model of costly signalling that was simpler and more accessible than Grafen's. It has since been extended by others (e.g. Johnstone and Grafen 1992; Bergstrom and Lachmann 1998) to cope with signals of graded intensity, error prone signals, and other problems. It provides a convenient general model of an 'action–response' interaction between a 'signaller' who may give one signal, and a 'receiver' who may make a single response, consisting of the transfer of a single resource. It can be thought of as a model of a chick begging from a parent, although it was not developed with that particular situation in mind. This context is simpler to analyse than the 'sexual selection' scenario considered by Grafen (1990*b*), because costs and benefits are easier to specify. The game was named after the poet-soldier Sir Philip Sidney (1554–86). His close friend Fulke Greville claimed that, fatally wounded in a skirmish near Zutphen, Sidney passed his water bottle to another casualty with the immortal words 'thy necessity is greater than mine'. Despite attracting scepticism, the story deserves to be true.

Since it is necessary to allow for a possible common interest between signaller and receiver, the model assumed that they are genetically related. However, as discussed in Section 2.6, there are other reasons why signaller and receiver could have a common interest.

Three versions of the model are described below. In the first, original version, signallers fall into one of two discrete classes, there is only a single kind of signal that can be given, and the receiver has only the choice between responding and not responding. In this discrete model, it turns out that not only there can be a 'signalling equilibrium' (i.e. an equilibrium in which signals are sometimes given and responded to) in which the signals are costly, but also that, depending on the values of the parameters, there can be an equilibrium with cost-free signals. This raises the question whether 'cost-free' signalling is a peculiarity of the discrete nature of the model. We, therefore, consider a second version in which we assume that the state of the signaller is a continuous variable, and a third version in which the intensity of the signal, and the response to it, are also continuous variables.

2.3.1 The discrete model

Signallers can be in one of two states, 'healthy' and 'in need'. If they are given the resource by the responder, they are certain to survive: if not, their chance of survival is $(1 - b)$ if healthy, and $(1 - a)$ if in need, where $a > b$. Potential donors are in only one state: if they transfer the resource in response to a signal, their chance of survival is $(1 - d)$: if they do not, they are certain to survive. (Note that if d varies between donors, the model is unaffected provided the variation in d does not alter behaviour: one just takes the average value of d). Summarizing, the probabilities of survival are

	Resource	
	Transferred	Not transferred
Potential donor	$1 - d$	1
Signaller		
In need	1	$1 - a$
Healthy	1	$1 - b$

If a signal is given, the signaller's chance of survival is reduced by c, the 'cost' of the signal, where $c \geqslant 0$.

Finally, the relatedness (i.e., the proportion of genes identical by descent between a signaller and a potential donor) is r. Both players act so as to maximize their own 'inclusive fitness' (Hamilton 1964), taken as equal to their own chance of survival, plus r times their partner's chance.

A signalling equilibrium is one in which the signaller only signals when in need, and the receiver only transfers the resource in response to a signal. It can easily be shown that the following conditions must be satisfied:

For the signaller: $a > c + rd$, or signaller never signals,

$b < c + rd$, or signaller always signals.

For the receiver: $a > d/r$, or receiver of signal never gives resource,

$b < d/r$, or receiver of signal always gives resource.

Several points are worth noting. First, the minimum cost of a signal needed to ensure that a healthy signaller does not signal, and thus, to keep the system honest, is $c = b - rd$. If $b < rd$, cost-free signalling is possible, because it would not pay a healthy signaller to signal, even if it was free.

It is interesting to compare this with Hamilton's (1964) famous conditions for the spread of a gene causing altruistic behaviour, $C < rB$, where C is the cost to the altruist, and B the benefit to the recipient. In the signalling case, b is the benefit to a healthy signaller, and d the cost to the donor. Thus, the condition $b < rd$ states 'honesty requires that the benefit of receiving a gift when healthy be less than r times the cost to the donor', whereas Hamilton's inequality states 'altruism requires that the cost of giving a gift be less than r multiplied by the benefit to the recipient'.

The second point is that the conditions for a signalling equilibrium are independent of the relative frequencies of healthy and in-need signallers. It follows that even very rare healthy signallers could impose the need for a costly signal on much commoner in-need signallers—a cost which would also lower the inclusive fitness of the receivers.

Finally, the conditions that must be met for a signal to be cost-free are equivalent to the requirement that signaller and receiver place the possible outcomes—resource given, or not given—in the same rank order; it is not necessary that they place the same values on the outcomes.

2.3.2 A model with continuously varying signallers

Suppose, now, that signallers do not fall into two discrete classes: instead, their chance of survival, $v = 1 - b$, varies uniformly between 0 and 1 (see Fig. 2.1). Otherwise, the model is the same as the one just analysed. Is cost-free signalling still possible?

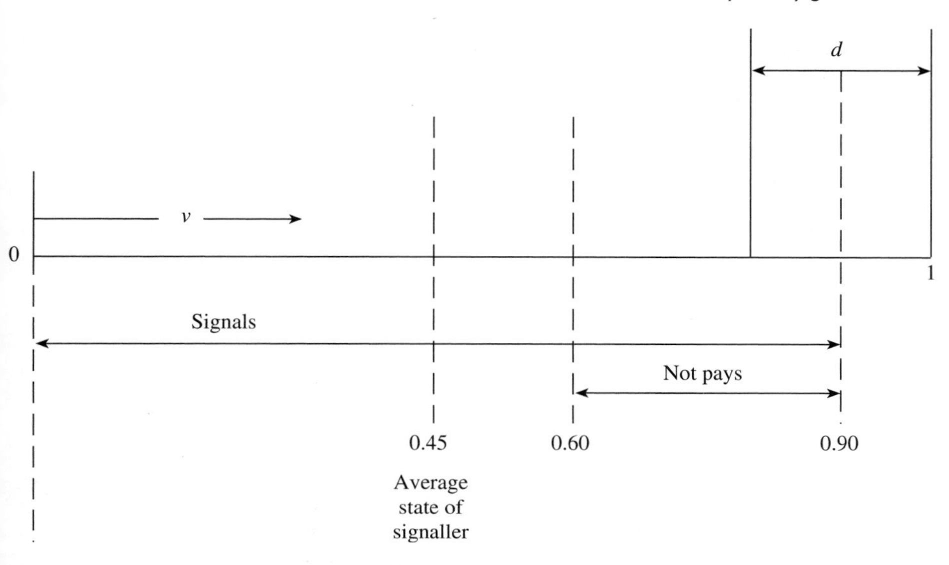

Fig. 2.1 A stable equilibrium with cost-free signals. The probability v that a potential signaller will survive without help is uniformly distributed between 0 and 1 in the population. The probability that a receiver will survive if he transfers the resource is uniform, and equal to $1 - d = 0.8$. Relatedness $r = 0.5$. It pays to signal if $v < v_{\text{crit}} = 0.9$. The average quality of individuals signalling is, therefore, 0.45. It pays the receiver to transfer the resource if $v < 0.6$, which on average is the case. However, in the region labelled 'not pays', the receiver would not transfer the resource if he knew the state of the signaller.

If a cost-free signal is to be stable, there must be some value of the signaller's condition, v_{crit}, above which it would not pay him to signal. That is

$$1 + r(1 - d) > v + r, \quad \text{or} \quad v < 1 - rd.$$

That is, if $v_{\text{crit}} = 1 - rd$, it will pay signallers to signal if their condition $v < v_{\text{crit}}$, but not otherwise. Suppose they behave in this way. Then, if a potential donor receives a signal, all he knows is that the condition of the signaller is less than v_{crit} and, since the distribution of v is uniform, that the signaller's average condition is $v_{\text{crit}}/2$. The receiver should, therefore, transfer the resource if

$$1 - d + r < 1 + r(v_{\text{crit}}/2), \quad \text{or} \quad v_{\text{crit}}/2 < 1 - d/r.$$

Figure 2.1 shows a numerical example in which a cost-free equilibrium is possible. For a lower value of relatedness, it would not pay the receiver to respond to a cost-free signal, and a signalling equilibrium would require a costly signal.

One point, illustrated in Fig. 2.1, is worth noting. Since the receiver knows only the average state of the signaller, and decides whether to transfer on that basis, he will be cheated some of the time: that is, he will transfer the resource although he would not do so if he knew the exact state of the signaller. This raises the question whether such exploitation is avoided if signals and responses can vary continuously.

2.3.3 A model with continuously varying signals and responses

Grafen's (1990*a*) model of Zahavi's idea allowed for continuous variation in both signal and response. We therefore modify the previous model by supposing that both the cost of the signal, and the magnitude of the response, can vary continuously. The only parameters that remain constant are the state of the receiver, $1 - d$, and the relatedness, r. This is equivalent to assuming that the response of the receiver does not vary with his condition, and is optimized for an average degree of relatedness.

Suppose that the cost of a signal, x, is a function $x = \varphi(v)$ of his condition, where $x >= 0$, and that the response is $y = \psi(x)$, where $0 \leq y \leq 1$: this implies a maximum possible transfer. Then the probability of survival of a signaller in condition v, who signals x and receives y, is

$$v + (1 - v)y - x.$$

Thus, a signaller who signals and receives the maximum transfer has a chance of survival of $1 - x$. The probability that a receiver transferring y will survive is $1 - dy^k$. Thus, as before, a receiver who does not transfer is sure to survive, and one who transfers the maximum survives with probability $1 - d^k$. The coefficient k allows the cost to the receiver of transferring part of the resource to vary non-linearly: if $k > 1$, transferring part of the resource is relatively cheap.

The problem then is to find two functions, $x = \varphi(v)$, and $y = \psi(x)$, such that, if the responder transfers according to $\psi(x)$, it would not pay the signaller to diverge from $\varphi(v)$, and vice versa. An analytical method for solving this kind of problem is described by Noldeke and Samuelson (1999).

We consider the case in which $d = 0.2$, and $k = 3$. That is, the chance of survival of a responder who gives y is $1 - 0.2y^3$, so that a responder who transfers the whole resource suffers a loss of fitness eight times as great as one who transfers only half the resource. This should favour a signalling equilibrium with graded responses, because we would expect circumstances in which it would pay the responder to transfer only part of the resource.

Figure 2.2 shows the solution when $r = 0.5$. As expected, there is an equilibrium with graded signals and responses, of the kind foreseen by Zahavi (1975) and Grafen (1990*a*). However, this is not the only possible outcome of an interaction in which graded signals and responses are possible. Thus consider the linear case, in which the receiver's chance of survival is $1 - 0.2y$. Then, if it pays the responder to transfer anything in response to a signal, it pays him to transfer the whole resource. It turns out that there are no continuous functions, $\varphi(v)$ and $\psi(x)$, satisfying the conditions outlined above. There can still be a signalling equilibrium, however, but

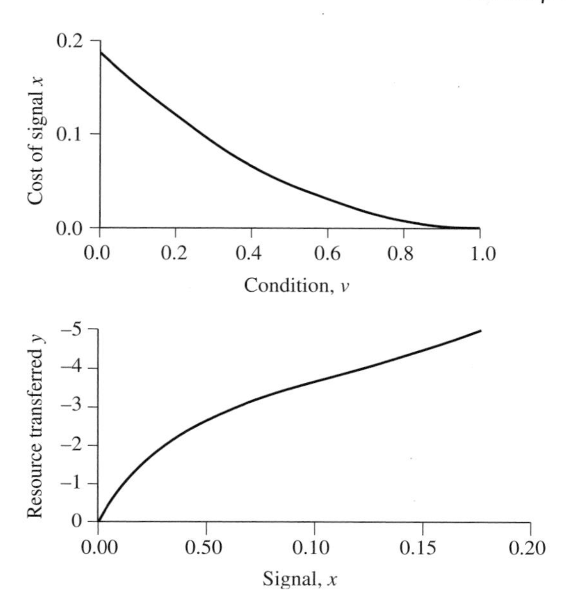

Fig. 2.2 An equilibrium with variable signals and responses. v; condition of signaller; assumed uniform between 0 and 1; x, cost of signal; y; proportion of resource transferred in response to a signal x. The assumptions of the model are as follows: relatedness $r = 0.5$; probability of survival of the signaller, $1 + (1 - v)y - x$; probability of survival of donor, $1 - 0.5y^2$. At the signalling equilibrium, the signaller chooses x as a function of v, and the donor chooses y as a function of x. Each actor chooses so as to maximize its inclusive fitness—that is, its own probability of survival plus r times its partner's.

one in which a signaller either gives a signal, or not, depending on his condition, and the responder transfers the whole resource in response to a signal. In other words, the problem is reduced to the one considered in the last section: there may be a signalling equilibrium with a fixed signal, which will be costly or cost-free depending on the value of r.

The models show that cost-free signals can be stable in a range of situations. They do not depend on the signals, or responses, being discrete. In the third model, the only parameter treated as constant is the state of the receiver of the signal. This assumption was made for simplicity: it is not needed for a cost-free signalling equilibrium. Bergstrom and Lachmann (1998) show that a cost-free equilibrium is possible even with a uniform distribution of the state of both signaller and receiver.

2.3.4 Perceptual error

The analysis in the last section suggests that at equilibrium the intensity of the signal will vary continuously. This conclusion conflicts with the observation, that in

many cases, an animal either does not signal at all, or signals with 'typical intensity' (e.g. Morris 1957; Cullen 1966). Johnstone (1994) has suggested that this discrepancy may arise because theoretical models usually ignore the existence of perceptual error: in the real world, recipients cannot perceive the advertising level of a signaller with complete accuracy. He analyses a model similar to that leading to Fig. 2.2, but incorporating perceptual error. Numerical simulations lead to the conclusion that, at an ESS, signalling is 'all or nothing': some proportion of the population, usually more than half, do not signal at all and the rest signal with uniform intensity. If perceptual errors are small, there may be more than one level of signalling, but the usual result of the simulations seems to be 'if of low quality, do not signal; if of high quality, signal with fixed intensity'.

An intuitive explanation of this result seems to be as follows. An individual whose quality is low, but not the lowest, gains little by signalling, since it may not be distinguished from the lowest quality individuals. Since signalling is costly, it is better not to signal at all, unless one can afford to give a signal indistinguishable from that of the highest quality individuals, and clearly distinguishable from the lowest. We accept that this intuitive explanation is not particularly convincing, but it is the best we can do.

Johnstone's conclusion is that the introduction of perceptual error does not alter the conclusion that the evolutionary stability of signalling may depend on the signals being costly to produce, but does explain why signals are often of typical intensity.

2.3.5 Conclusions

Analysis of the Philip Sidney game confirms Grafen's (1990*a*) conclusion that a signalling equilibrium can exist, with the following characteristics:

1. Signals accurately reflect the state of the signaller (honesty).
2. Signals are costly to make.

Grafen added a third requirement, that a given signal is more costly to make for a signaller of lower quality. In the model just analysed, we assumed that the cost of a signal is independent of the quality of the male, but the benefit of a given response is greater for signallers of lower quality. Further confusion may arise because in Grafen's model males of high quality give more costly signals, whereas the reverse is true in the Philip Sidney game. However, there is no real contradiction. What is needed is a more general formulation of Grafen's third condition, as follows:

3. The ratio of the cost of a signal to the benefit received is lower for individuals giving stronger signals. The following table illustrates this third condition for the

two models:

Model	Signal strength	Quality of signaller	Benefit b	Cost c	Ratio c/b
Grafen	High	High	1	Low	Low
	Low	Low	1	High	High
Philip Sidney	High	Low	High	1	Low
	Low	High	Low	1	High

However, analysis of the Philip Sidney game has introduced some additional possibilities. In particular, it has shown that there may be signalling equilibria which are honest, but in which signals are no more costly than is needed to convey the information: that is, there is an efficacy cost, but no strategic cost. This possibility arises because signaller and receiver have a common interest. What is required is that they should place the possible outcomes of the interaction in the same rank order of preference: they may, and typically will, differ in the strength of their preferences.

In the Philip Sidney game, a common interest is possible because it is assumed that signaller and receiver are related. This is true, for example, for signals between parents and offspring, and between members of eusocial insect colonies. There are, however, other contexts in which signaller and receiver have a common interest. This will often be true for signals between mates or among members of a social group, even if they are not relatives.

2.4 'Pooling equilibria'—a more radical proposal

Bergstrom and Lachmann (1998) propose the notion of a 'pooling equilibrium' which presents a serious challenge to the handicap model. We will illustrate their idea in the context of a chick begging from a parent. The model they consider is as follows. Chicks vary continuously in 'need', and parents vary continuously in the cost, to them, of feeding. So (allowing for relatedness, and calculating 'inclusive fitness'), there are three possibilities for a particular parent and chick:

(1) feeding pays both chick and parent;
(2) feeding pays neither chick nor parent; and
(3) there is a region of conflict: the chick would benefit from being fed, but the parent would benefit from not feeding.

Now suppose that a chick can either not signal at all, or give a cost-free signal: the crucial assumption is that only one level of signal is possible. At the ESS, chicks signal only if they are above a threshold level of need. Hence, a parent knows only the *average* state of a signalling chick, and can respond only to that average: sometimes, from her own point of view, she will give when she should not, or *vice versa.* Bergstrom and Lachmann show that an ESS is possible. At the ESS, no signalling costs are paid, and both chick and parent are better off than they would be if signals could vary continuously. In the latter case, an ESS would be reached (see Section 2.3.3), with costly signals.

The result seems paradoxical. Both chick and parent are better off than they would be if complete information about state could be transmitted (but not as well off as they would be if complete information could be transmitted cost-free). But there is no doubt that the conclusion is mathematically correct, given the assumptions. The authors extend the model to cases in which more than one level of cost-free signal is possible: for example, no signal, signal A, signal B, and signal C. They show that, at the ESS, chicks fall into four groups, or 'pools', according to need, and that parents would likewise fall into four groups.

The possibility of 'pooling equilibria' with cost-free signals presents a possible alternative to the handicap model. Before considering the empirical evidence, there is a possible difficulty with pooling equilibria to be discussed. Suppose there are only two pools, signal and no signal (the difficulty is more severe if there are several levels of signal). In practice, there will be some genetic variation in behaviour. The ESS would be stable against mutations affecting readiness to signal, or to respond, as a function of state: that is what an ESS means. But the non-variability of signals is an assumption, not a conclusion from the model. Everything depends on the plausibility of this assumption. Suppose that the intensity of the signals varies. In particular, suppose that

(1) the intensity of the signal is (sometimes) a bit higher in chicks that are very hungry, and a bit lower in chicks only just hungry enough to beg;
(2) parents close to the threshold for response are a bit more likely to respond to a higher intensity signal, and less likely to respond to a lower intensity signal: parents well above the threshold respond anyway.

If these assumptions are true, then the pooling equilibrium will be unstable. A 'mutant' chick signalling more strongly when very hungry will have the same pay-off as a typical chick when interacting with a typical mother, and will do better when interacting with a 'mutant' mother that is more likely to respond to a slightly stronger signal, and a similar argument shows that 'mutant' mothers also have a selective advantage. The fixed intensity signal would evolve into one that varied with need. The end result would be a system of continuously varying and costly signals.

The crucial question, of course, is whether these assumptions about variation are plausible. There do not have to be specific 'mutants'. It is sufficient that, for mechanistic rather than adaptive reasons, the intensity of the signal increases slightly with need, and that readiness to respond declines slightly as the state of the responder falls towards the response threshold. These assumptions seem plausible, although empirical support, or refutation, would be welcome. If the assumptions are correct, then cost-free 'pooling equilibria' will be unstable. We return to the problem in Section 3.2.3.

2.5 Non-signalling equilibria

It is important to remember that, in addition to the 'signalling equilibria' discussed above, a game such as the Philip Sidney game will also have a 'non-signalling equilibrium'—either 'never signal, never give' or 'never signal, always give', depending on whether in the absence of information the potential donor benefits by not giving, or by giving. The population may evolve to the signalling or to the non-signalling equilibrium, depending on initial conditions. How, then, can signalling ever evolve, if the initial state is not to signal? After all, why should the first individual to signal be taken notice of? A likely answer is that, before signals evolve, potential donors base their decisions on cues (which have not evolved), and that these cues evolve into signals by a process that has been called ritualization (see Section 5.1).

Bergstrom and Lachmann (1997) have pointed out that the average pay-off to both signaller and receiver may be greater at the non-signalling than at the signalling equilibrium. It is useful to have an intuitive feel for why this is so. Imagine that a chick, which varies in degree of hunger, signals to its mother. Suppose that usually the chick is in serious need, but that sometimes it is not. If there are no signals, it pays the mother always to feed the chick: it costs her something when the chick is not hungry, but this is more than counterbalanced by the typical cases in which she feeds a hungry chick. So 'never signal, always feed' is an ESS. Since there is no signalling, neither partner pays the costs (direct for the chick, and in inclusive fitness for the mother) of signalling. Hence both are better off than they would be at the signalling ESS.

2.6 Must honest signals always be costly?

Analysis of the Philip Sidney game has shown that the answer to this question is in the negative. In particular, it has been shown that signals can be both reliable and cost-free if signaller and receiver place the possible outcomes of the interaction in the same rank order of preference: they may, and typically will, differ in the strength of their preferences.

The two most familiar and widely accepted reasons why two animals might prefer the same outcome to an interaction are relatedness and reciprocation (Trivers 1971). However, Zahavi (1977, 1995) rejected both these explanations, and suggested that 'altruism' was only apparent, arguing that altruists improved their future fitness by signalling their 'social prestige' to prospective mates or allies. The case for this idea is perhaps strongest when the receiver of the signal is the sole recipient of the altruism, for two reasons. First, some of the costs might be recouped; for example, 'courtship feeding' by male Ospreys increases female breeding success as well as allowing females to assess male foraging ability (Green and Krebs 1995). Second, receivers ought to be able to make unusually accurate and efficient estimates of signalling effort. Despite these advantages, however, it is hard to accept Zahavi's explanation for apparent altruism in competitive situations, if the receiver of the signal benefits at the expanse of the signaller (Wright 1999). Even if the altruism benefits third parties, which would avoid this difficulty, signallers would still be vulnerable to 'cheats' eliciting extra benefits. Zahavi (1995) attempts to get round this difficulty by his concept of 'social prestige', which is not only increased by altruistic acts, but also decreased by soliciting or receiving help. Thus it is not the same as dominance rank (Zahavi 1990). We are, however, uncertain exactly what social prestige is and how to measure it; like Wright (1999) we find it hard to disentangle from well-studied concepts such as reciprocity and alliance-formation. Moreover, cooperative breeding by Arabian Babblers, the behaviour that Zahavi uses to illustrate his ideas, seems fully consistent with orthodox explanations of altruism (Wright 1997).

There are other situations in which cost-free signals can be reliable. These are listed briefly here, and will be discussed in detail in later chapters.

1. *Indices.* A signal may be honest because it cannot be faked. Such signals are the topic of Chapter 4.
2. *Coordination games.* Signaller and receiver prefer different outcomes, but share an overriding common interest—for example, in avoiding an escalated contest or the break up of an alliance of value to both participants. The use of cost-free signals to settle such interactions usually depends on the recognition of some asymmetry—for example, in age, sex or ownership of a resource—by both partners. Such games are discussed in Section 3.3.
3. *Repeated interactions.* The role of signals in such cases is discussed in Section 7.5.
4. *Punishment of false signals.* This is discussed in Sections 6.4 and 7.6.2.

2.7 Conclusions

This chapter has been primarily concerned with the idea, originating with Zahavi (1975), that the reliability of signals depends on their cost. First, it is essential to distinguish between 'strategic costs' that are necessary in order to ensure reliability and prevent 'lying', and 'efficacy costs' that are required to convey information unambiguously, even when the signaller has no inducement to lie. Theoretical analysis of strategic costs (in particular, Enquist 1985; Pomiankowski 1987; and Grafen 1990*a*,*b*) has shown that there are indeed contexts in which the reliability of signals can be ensured by costs. In addition, their analysis showed that stability requires that the cost of a given signal must be relatively greater in signallers of low 'quality': a more general formulation of this requirement is that the ratio of the cost of a signal and the benefit received must be greater in individuals giving stronger signals, either because their high quality enables them to produce a signal at lower cost in fitness, or because their need for a response is greater.

A convenient model to analyse these problems is the Philip Sidney game (Maynard Smith 1991). The essential features if this model are, first, that it is a simple 'Action–Response' game with a single signal and a single response, and second, that signaller and receiver have some shared interest: this is ensured by assuming that they are related and that each acts so as to maximize its inclusive fitness, but other sources of common interest are possible. The main conclusions are as follows:

1. There are contexts in which only costly signals are reliable.
2. There are also contexts in which cost-free signals can be reliable: this is the case if signaller and receiver place the possible outcomes of the interaction in the same rank order of preference.
3. The requirement for evolutionary stability of costly signalling is that the fitness cost of a given signal must be relatively lower for signallers of high 'quality': a more general formulation of this requirement is that the ratio of the fitness cost of a signal and the benefit received must be lower in individuals giving stronger signals, either because the signal is less costly for a signaller of high quality, or because individuals give stronger signals when in greater need.
4. The main conclusions continue to hold if signallers and receivers vary continuously in quality, and if variable responses are possible.
5. If there are errors in the perception of the signal, this may lead to an equilibrium in which a continuously varying signal is replaced by discrete signals of uniform intensity.
6. In addition to an equilibrium in which signals are given and responded to, there will be a non-signalling equilibrium—either 'never signal, never respond', or 'never signal, always respond'. The average pay-off to both signaller and receiver may be higher at the non-signalling than at the signalling equilibrium.
7. A radical alternative to the above analysis has been suggested by Bergstrom and Lachmann (1998). If it is assumed that only a few discrete non-varying signals are possible, there may be a 'pooling equilibrium', in which signals are cost-free, and 'pools' of similar signallers all give identical signals, and receive identical responses. We point to some difficulties with the assumption of non-varying signals, but the idea deserves further study.

In Chapter 3, we discuss some specific cases that can plausibly be interpreted as examples of signals stabilized by strategic costs, or as examples of cost-free signals between individuals with identical preferences. In later chapters, we discuss other alternatives to the idea that reliable signals must be stabilized by strategic costs—in particular, unfakeable indices, punishment, coordination games, and, in social animals, reputation.

3

Strategic signals and minimal-cost signals

3.1 Introduction

The first two chapters have been primarily concerned with the evolutionary processes that ensure the reliability of animal signals, despite the potential advantages of lying. Our aim has been to emphasize that there is no single answer to this problem. In this chapter and the next, we pursue the matter further, by suggesting a number of specific examples of signals that fall into one or other category, and also some examples whose explanation seems to us doubtful.

The three main reasons for reliability that we have suggested are the following:

1. The 'handicap principle': the signal is costly, and reliable because it is too costly for a signaller of low quality. It may be costly to produce, or it may have costly consequences, perhaps because it incurs risks of retaliation. The theory of such costly signals was outlined in the last chapter.
2. The signaller would not gain by lying, even if the signal was cost-free. The obvious example is that signaller and receiver have a 'common interest'—that is, they rank the possible outcomes in the same order of preference: this case was discussed briefly on p. 26. There are, however, a number of other situations, discussed in Section 3.3, in which the signaller would not gain by lying.
3. The signal cannot be faked.

In this chapter we discuss some cases which, we think, are best interpreted either as examples of the handicap principle, or of common interest. The third category, which we have called indices, are discussed in the next chapter. Cases, other than 'common interest', in which it would not pay the signaller to give a dishonest signal are discussed in more detail in Chapters 6 and 7.

It is hard to see how one could estimate the relative importance of the handicap principle and of common interest. There is, however, one reason why biologists may have been led to overestimate the importance of handicaps. It is not surprising that studies have concentrated on dramatic, costly signals like the song of the Nightingale and the tail of the Peacock. But this should not blind us to the fact that many signals are almost furtive, a point emphasized by a recent introduction to animal communication, entitled *Not Only Roars and Rituals* (Rogers and Kaplan 1998). For example, recordings made from birds when close to their mates often reveal a veritable stream of calls so quiet that they are inaudible to humans more than a few metres away (e.g. European Robin, Cramp 1985).

3.2 Strategic signals

How common is handicap signalling; that is, signalling that is reliable because it is costly? In answering this question, two points have to be borne in mind. First, it is not sufficient to show that a signal conveys reliable information despite the absence of a common interest between signaller and receiver. The signal may be just an index, not itself costly, of some quality of interest to the receiver: such signals are the subject of the next chapter. To demonstrate the existence of a handicap, it must be shown that the signal could vary independently of the quality signalled, and hence be unreliable: if it does not do so, it is because it would not pay the signaller to make a dishonest signal, and this in turn implies that the signal must have a strategic cost. Second, the theory of strategic signalling suggests that the ratio of costs to benefits must be lower for those individuals that signal most strongly: this may be difficult to demonstrate in many cases.

3.2.1 Stalk-eyed flies

We start with an example in which the evidence for handicap signalling is convincing. In stalk-eyed flies, *Cyrtodiopsis dalmanni*, males have long eye stalks (Fig. 3.1). It has been shown that females prefer to mate with males with a large eyespan (Wilkinson

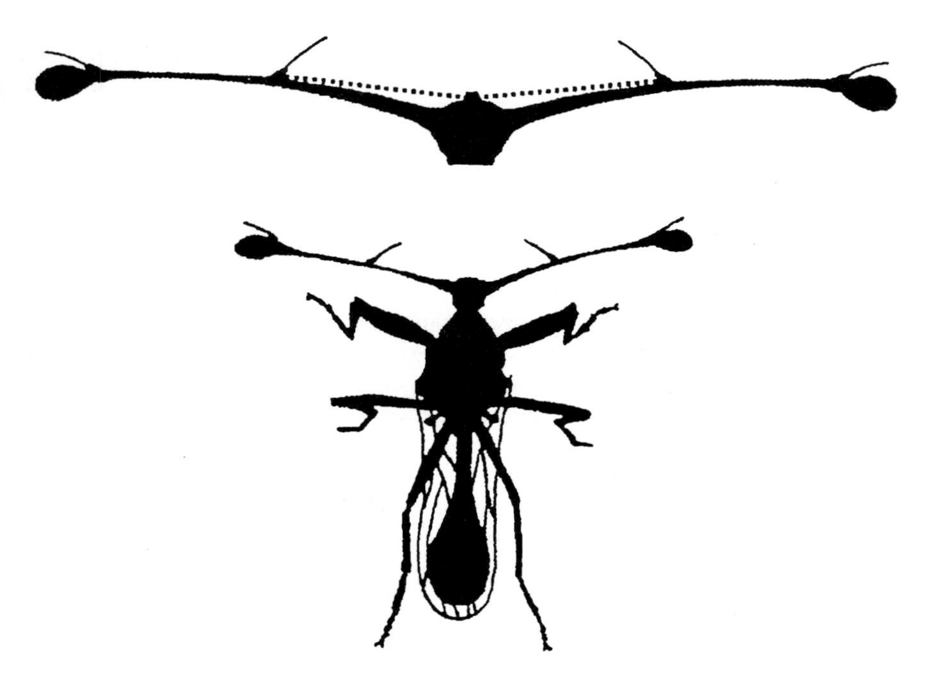

Fig. 3.1 A male stalk-eyed fly (from David *et al.* 1998).

and Reillo 1994); the long eye stalks are certainly a 'signal' according to our definition. David *et al.* (2000) raised larvae belonging to different families on three diets (corn, spinach, and cotton wool), and measured eye span, thorax, and wing length in adults of both sexes. Their crucial findings were as follows:

1. Male eye span varied more with level of nutrition than did female eye span, or non-sexual traits. This phenomenon is known as 'condition-dependence' of sexually selected traits: similar results have been observed in other organisms (e.g. Hill 1992, in House Finches; Hunt and Simmons 1997, in horned beetles *Onthophagus taurus*; Gustaffson *et al.* 1995, in Collared Flycatchers).
2. There was genetic variance for the degree of condition dependence: some genotypes retained large eye spans on all three diets, whereas others had drastically reduced male eye spans on the poorest diet.

These findings can best be explained if high-quality males allocate more resources to a sexual signal than do low-quality males, at the same cost. Note that, if female flies were getting parental care from their mate, it would pay them to select males with a larger eye span, even if differences were entirely environmental in origin. In fact they are only getting 'good genes'. However, since the ability to produce a large signal despite a poor environment is in part genetic, it still pays females to select such males.

In this example, then, there is evidence both that the cost of a given signal is relatively lower for high-quality males, and that there is genetic variance for the strength of the signal, and its dependence on quality. The conditions for the evolution of a handicap equilibrium are therefore present.

3.2.2 Musth in elephants

Another case that is probably best interpreted as a handicap is the phenomenon of musth in elephants: the account that follows is based on Poole (1989). Male African Elephants are sexually active for only part of the year. When active, they move from group to group seeking females in oestrus. For part of the period of sexual activity, a male is in musth, a state of aggressiveness that is signalled by secretions from the temporal gland, dribbling of urine, and by vocalization and posture; the circulating testosterone level is raised by a factor of approximately 50. Older and larger males remain in musth for 2–5 months, usually in the wet season when more females come into oestrus; younger males are in musth for from several days up to a few weeks, usually in the dry season.

Agonistic interactions between males in the same state, in musth or not, are usually won by the larger. Interactions between a male in musth and one not in musth are usually won by the former: Poole observed that contests in which only the smaller male was in musth were won by that male in 86% of cases. If both contestants were in musth, the smaller male retreated, and the symptoms of musth were often suppressed: urine dribbling ceased, temporal gland secretion dried up, and posture changed. Escalated fights were very rare. Only 31 such fights were observed in 14 years of study. Of these, 8 were between a musth and a non-musth male. The smaller male was in musth in seven out of eight cases, and won the fight in six cases. Twenty fights were between musth males of approximately equal size; two of these fights ended in the death of one male.

These observations suggest that musth is a signal of intention to fight, and is respected because of the risks of an escalated fight. Why, then, do males not cheat? Smaller males can be in musth, and can then win contests against a larger opponent. Maynard Smith (1982) suggested that the explanation might be that a male in musth is irrevocably committed to escalation. If so, it would explain why a larger non-musth male retreats. If musth is rare enough, the cost of escalated fights on the rare occasions on which two musth males meet would be balanced by the gain on the much commoner occasions on which only one male is in musth. Poole's observations show that Maynard Smith's proposed explanation is wrong. Musth is not an irrevocable commitment: a smaller male in musth retreats without escalation if its larger opponent is also in musth.

As Poole suggests, the correct explanation probably depends on the fact that being in musth has costs additional to those arising from the risks of escalated fighting. Males in musth were observed to lose condition, probably because of increased activity, greater metabolic rate resulting from high testosterone levels, and loss of liquid by urination. It seems certain that these costs would prevent any male being continuously in musth. It is also likely that these costs are greater for smaller males, which have more to lose by a loss of condition which would inhibit growth into a larger and more successful male. If it is true that musth has metabolic costs, and that the signalling of musth is impossible without paying those costs, the system can be explained. It will pay larger males to come into musth in the wet season, when more females are in oestrus. Smaller males will then avoid the wet season, since they would lose contests against larger musth males, and their musth will be confined to the less profitable dry season. If this interpretation is correct, musth is an example of Zahavi's handicap principle.

3.2.3 Chick begging

Chicks often beg for food in ways that appear costly, both energetically and in predation risk. Today, the most widely accepted explanation is that the signals are a handicap: they provide information to the parent about the chick's state, and are costly because only then can they be reliable (Godfray 1991; Godfray and Johnstone 2000). We discuss the empirical support for this view later, but first we list some alternative theories, because only then can the relevance of the empirical data be appreciated.

1. *Sensory exploitation.* Chicks are exploiting an innate parental response. This was the common view among early ethologists familiar with the idea of a supernormal stimulus. Trivers (1974) pointed out that there may be a conflict of interest between parent and offspring over the amount of food transferred. Sensory exploitation cannot be a complete explanation, because parental responses can evolve and, since the parent is in charge of the exchange, will do so until parental exploitation is minimized.

However, the possibility of sensory bias should not be ignored: we return to this point below when discussing 'pooling equilibria'.

2. *Sibling competition.* A chick must make costly signals in order to compete with its siblings for a limited supply of food. Such competition must surely occur, but is difficult to model, and empirical tests of its effects are hard to design and interpret (Godfray 1995; Godfray and Parker 1992). In practice, the effects of competition and of selection for reliability are likely to operate alongside one another in most cases. They will have similar consequences: signals will be costly, both because they must compete, and because they must be reliable.

3. *Pooling equilibria.* In Section 2.4, we described a model proposed by Bergstrom and Lachmann (1998) suggesting that reliable information can be conveyed by two (or more) distinct signals, with individuals in a particular range, or 'pool', using the same signal—for example, individuals above a certain threshold signalling A, and those below signalling B. However, it is an assumption of their model that there can be no perceptible variation in the intensity of a particular signal, for example A. We argued that this assumption may be incorrect, and that if it is relaxed the stability of the signalling equilibrium disappears. However, we see this as a reason for being cautious about the idea, rather than for outright rejection.

It has been suggested (Godfray and Johnstone 2000) that this model might offer an alternative to the handicap model as an explanation for the variation in chick begging signals. The two models should be distinguishable empirically by two criteria. If the handicap model is correct:

(1) signals should be costly; and
(2) signals should be more intense for individuals in greater need.

Neither of these effects are expected on the 'pooling equilibrium' hypothesis.

Given these clear predictions, one might suppose that the issue could be settled empirically. Consider first the costs arising from the risk of predation: in effect, the chick is blackmailing its parent into feeding it, because the parent has a strong interest in the chick's survival. It seems plausible that loud calls should attract predators: every bird-watcher knows that a good way to find a nest is to listen for begging calls. There is empirical evidence (Haskell 1994) that calls do attract predators. But it is harder to provide evidence for a relation between cost and need.

The energetic costs of begging are more controversial. Several authors (e.g. McCarty 1996) have found that the metabolic rate of a calling chick is only some 1.25 times that of a resting chick, and have argued that the energetic cost of begging is small. Kilner (2001) has criticized this conclusion on the grounds that, in these experiments, chicks were allowed to choose their own level of calling. She estimated the cost of calling in Canary chicks by comparing pairs of siblings forced to beg for 10 s and for 60 s (periods that are within the natural range) before being fed, and measuring the decline in weight gain so caused. The chicks forced to beg for a longer period showed a significantly greater weight loss at age 8 days, but not at 6 or 10 days. Kilner argues that this supports the view that signals are costly, because 8 days corresponds to the age of maximum growth rate, and therefore, is likely to be the most sensitive age. Her results must be taken seriously, but do not seem decisive: for example, at age 6 days the *relative* rate of weight gain appears to have been higher than at 8 days, yet the chicks showed no increase in weight loss when forced to beg for longer.

Evidence on the second point, that signals are more intense in chicks in greater need, is more convincing: Kilner and Johnstone (1997) quote a number of examples (e.g. Redondo and Castro 1992, on magpie chicks).

To summarize, a number of alternative theories have been proposed to account for chick begging signals. There is no reason to suppose that one of these theories is correct for all species, or that only one mechanism is operating in any given species. In the current debate between proponents of 'handicap' theories, and cost-free theories depending on 'pooling equilibria', the evidence that signals are more intense in chicks in greater need, and the fact that pooling equilibria require the existence of a series of distinct but invariant signals, leads us to prefer the handicap model. But we do not regard the issue as settled.

3.3 Minimal-cost signals

3.3.1 When can minimal-cost signals be evolutionarily stable?

By a minimal signal, we mean a signal whose reliability does not depend on its cost (i.e. excluding handicaps), and which could be given by any signaller (i.e. not an index). There are four contexts in which such signals can be evolutionarily stable.

1. Signaller and receiver place the possible outcomes of the interaction in the same order of preference, although they may differ in the value they place on different outcomes (see p. 26): we have called this 'common interest'.

2. Dishonest signals are punished (Maynard Smith and Harper 1988; Viljugrein 1997). Lachmann *et al.* (2000), in a paper entitled 'The death of costly signalling', accept that handicaps—that is, signalling systems maintained because honest signals are costly—will exist, but point out that, if honest signals are cost-free and only dishonest signals are costly, then at an evolutionary equilibrium signalling will be cost-free. They suggest that such a situation will exist when there is socially imposed punishment of cheaters. They mention human language as an example of a cost-free signalling system in which honesty is maintained by a social sanction against liars. In Chapter 6 we discuss cases in which punishment of lying may be important in animal signalling systems. The difficulty, of course, is not to explain how the threat of punishment can maintain the honesty of a cost-free signal, but why punishing behaviour evolves if it is potentially costly to the punisher.

3. Signaller and receiver prefer different outcomes, but have a strong common interest, for example, in avoiding an escalated fight or the break up of a mated pair. Such interactions are referred to in economics as 'coordination games' (see Silk *et al.* 2000). The classic example is the 'War of the Sexes' game. Two people are trying to arrange an outing: Pat wants to go rock-climbing, but Chris wants to go the zoo.

Both, however, would prefer to be together than pursue their preferred activity on their own. Two humans would settle such a disagreement by discussion, perhaps tossing a coin to decide. Animals cannot do this, but they may use a pre-existing asymmetry to settle the outcome. For example, the Hawk–Dove game is a coordination game, in the sense that both contestants have an overriding interest in avoiding an escalated contest. Maynard Smith and Price (1973) showed that an ESS of the game is the Bourgeois strategy, 'if owner, play Hawk; if intruder, play Dove'; here, 'owner' means, roughly, 'have been in occupation for some time'.

It is worth thinking more carefully about how the Bourgeois strategy could be realized in practice. Maynard Smith and Price assumed that every contest was between an 'owner' and an 'intruder', that the participants knew their own role before the contest began, and that a contest between two Doves could be settled without cost. Consider a territorial contest over an area too small to be shared. What a contestant would 'know' is how long it had occupied the territory. Typically, a contest would be between an individual that had been in occupation for some time (the 'owner') and one that had only just entered (the 'intruder'). Such contests could be settled as Maynard Smith and Price imagined, by following the rule 'display for a length of time that increases with the time you have occupied the area, and then retreat'. But, at the start of the season, some contests will be between two intruders. Clearly, it will pay an intruder to display for some time, at some cost, before retreating. Hence, contests will never be completely cost-free.

Contests over indivisible resources are discussed in some detail in Chapter 6. The immediate point, however, is that a coordination game, in which the contestants prefer different outcomes but share a common interest in avoiding escalation, can be settled by an asymmetry, but only if the asymmetry is known unambiguously to both contestants. Can signals help to settle such contests by conveying reliable information about asymmetries? Clearly, the answer is yes, but only if the signal is unfakeable—that is, it would have to be an index. In fact, many of the indices discussed in Chapter 4 (in particular, indices of size, ownership, and condition) act to reveal asymmetries.

4. Repeated interactions take place between the same pair of individuals. Silk *et al.* (2000) show that, if a pair interact repeatedly, an ESS with minimal-cost signals may be possible. Such an ESS requires that the receiver of signals should recognize and remember individuals; in other words, it depends on what we have called 'reputation'. The role of repeated interactions is discussed in Chapter 7.

Summarizing, minimal-cost signals can be evolutionarily stable if

1. Signaller and receiver place the possible outcomes in the same rank order.
2. Dishonest signallers are punished.
3. Although they prefer different outcomes, signaller and receiver also have an overriding common interest. The outcome of such interactions may be settled by a pre-existing asymmetry, for example in ownership, which may be recognized through minimal-cost signals.

4. In the absence of a suitable asymmetry, minimal-cost signals may settle such coordination games if the same pair of individuals interact repeatedly.

We now describe examples of the first category; 'common interest': the roles of asymmetries, of punishment, and of repeated interactions are discussed in later chapters.

3.3.2 Signals between unrelated individuals with a common interest

Female fruit flies, *Drosophila subobscura*, mate only once in their lifetime, and store sperm sufficient to fertilize all the eggs they will lay. If a male is placed with a mated female, she extrudes her ovipositor towards him, and he ceases courtship immediately. In the absence of such a signal, a male will continue courtship for up to an hour, even if his advances are rejected (Maynard Smith 1956). Forced copulation is impossible; it, therefore, pays the male to pay immediate attention to the signal; it also pays the female to bring to an end the male's interference. The cost of the signal in time and energy is negligible. Both male and female benefit from the cessation of courtship, brought about be a minimal-cost signal.

Dawkins and Guilford (1994) describe a similar case in the Bluehead Wrasse. Males defend their territories against intruders, and are unable to mate while doing so. To indicate readiness for mating, a male changes colour from green to opalescent grey, and swims in tight circles above the female. In response, the female swims head up towards to the surface, initiating the synchronous release of gametes. Since both male and female benefit from this synchrony, there is no conflict of interest between them, so the signals exchanged need not be costly, although they are certainly more elaborate than those that cause a cessation of courtship in *D. subobscura.*

Turning to a quite different context, a variety of animals signal to attract conspecifics to food (Wrangham 1986; Brown *et al.* 1991). The advantage to the receiver of such a signal is obvious, but what is in it for the signaller? Why attract competitors? The answer seems to be that feeding alone has the disadvantage of higher predation risk compared with foraging in a group.

An interesting example is provided by a study of House Sparrows (Elgar 1986). Individuals that found food often gave *chirrup* calls. When they did so at a high rate, other sparrows joined them more rapidly than when they were less noisy. Playback confirmed that the calls alone could attract other sparrows. The rate of *chirrup* calls depended on two factors. First, sparrows called more when the food was easier to share, as when bread was presented as crumbs rather than a single lump (when aggression was common). Second, call rate was lower in larger groups, suggesting that the net benefit of calling fell with increased competition. This study raises some additional questions:

1. Are other species attracted by *chirrup* calls?
2. Does apparent risk of predation alter calling rate?
3. Dividing bread up into crumbs might increase the caller's perception of food quantity rather than the ease with which it can be shared (Hauser 1996).

Minimal-cost signals will be favoured whenever the intended receivers are close by and when the costs of being detected by eavesdroppers are high. Two types of contact calls can be heard from winter flocks of Crested Tits: a loud purring tremolo and a weaker higher pitched *zizi* audible only at close range (Cramp and Perrins 1993). Stuffed birds were attacked significantly more often by sparrowhawks if accompanied by playback of the long-range contact calls than when the tape recorder played short-range calls (Krams 2001). Unfortunately for them, Crested Tits often need to make loud calls: for example, they roost alone in holes scattered across a group winter territory of about 20 hectares and so need to make contact with each other every morning. Once the flock has reformed, however, it is no surprise that most calls are quiet.

Killer Whales, when hunting Elephant Seals, use very few contact calls and these are unusually quiet. In this case not only is there the risk of the seal detecting the signals, but signalling may interfere with the whales' attempts to hear their prey. Once a seal has been caught, however, the whales give many loud contact calls which can attract other group members from several kilometres away (Guinet 1992).

Minimal-cost signals also occur between members of different species: perhaps the clearest example is the use of warning colouration by distasteful or toxic prey to deter predators (Fig. 3.2). There is a clear common interest. However, the signals

Fig. 3.2 Skunk warning signal (from McDonald 1984, vol. 1).

may be costly to produce, and may make the prey individual more easily seen by specialized predators. Bees and wasps signal by their coloration that they are potentially dangerous, but there are predators that can disarm their stings. However, as Guilford and Dawkins (1991) argue, the bright colouration may have evolved because such patterns are more easily learnt and remembered by predators, and not because they must be costly in order to be honest; the cost is an efficacy cost, not a strategic one.

A fascinating example, involving signals by each of two cooperating species, has been described by Isack and Reyer (1989). Humans have collected honey in Africa for 20,000 years. Honeyguides guide humans to colonies of honeybees: the humans obtain honey and the birds feed on the larvae and wax from the disrupted nests. The Boran people of Kenya attract the honeyguides by a loud whistle, audible up to 1 km, produced by blowing into their fists or into modified snail shells or palm nuts. The birds guide the search by indicating both the direction and distance of the nest, reducing the average search time from 8.9 to 3.2 h. The information is conveyed by a short undulating flight from perch to perch, during which the birds display their white outer tail feathers and give a characteristic call. The authors confirmed that the information about distance and direction the Boran people claimed to obtain was indeed present in the birds' signals. The original interaction of honeyguides may have been with Honey Badgers, which they also guide to bee colonies.

3.3.3 Relatedness

The classic example of honest communication between relatives occurs in social insects. Proper functioning of the colony requires that workers perform the correct tasks—for example, foraging, feeding the brood, cleaning the nest—in the correct way. This requires the exchange of signals. In honey bees, Seeley (1998) lists 17 different types of signal, chemical or mechanical, exchanged between members of a colony. For example, there are signals given by workers that concern brood recognition, eliciting of aggression and recruitment of food sources, and by the queen that indicate the queen's presence or that an egg was laid by the queen (workers may lay unfertilized eggs which develop as males unless killed by other workers). Colony integration requires that bees respond not only to signals but also to cues, and their response to a signal may depend on such cues. For example, the waggle dance indicates not only the direction and distance of a food source, but also its nature—pollen, nectar or water; foraging workers respond differently depending on the current level of these three resources. Seeley lists over 30 cues influencing the behaviour of workers. He suggests that the excess in the number of cues over that of signals reflects their relative importance in ensuring appropriate behaviour.

It is natural to compare the cooperation between workers to produce a functioning colony, and between the cells of the body to produce a functioning organism. However, it should not be pushed too far. The workers in a colony, although related, are not genetically identical, so conflicts arise (Ratnieks and Reeve 1992). In primitively eusocial hymenoptera, sibling females fight over which will become a queen and which workers. In more advanced hymenoptera, a worker can do more to transmit her own genes by raising her own unfertilized eggs than by raising eggs laid by the queen, so that policing by other workers has evolved. Nevertheless, the possibilities of direct reproduction open to workers are limited, and this has resulted in a high degree of cooperation. Workers signal to one another just as the cells in the body do, and there is no reason (despite Zahavi and Zahavi 1997) to think that these signals between cells have to be costly in order to be reliable. If two cells are genetically identical, the circumstances in which the signalling cell is selected to elicit a response are identical to those in which the receiving cell is selected to respond. Things are not so simple in social insects, because signaller and receiver are not genetically identical, but the analogy is helpful if used with caution.

3.3.4 Kin recognition

There are many contexts in which animals behave differently to kin and non-kin. This may depend merely on familiarity: for example, an animal that avoids mating with a littermate is avoiding mating with a close relative. But, in some cases, an animal is able to treat unfamiliar conspecifics differently according to relatedness. This requires some transfer of information. The topic is reviewed by Sherman *et al.* (1997). Recognition requires that an animal compare information (which Sherman *et al.* refer to as a 'label') concerning another individual with some internal representation, or 'template', of their own genotype. The topic is complex, both because of the variety of contexts in which kin recognition occurs, and the different ways in which labels and templates are constructed. For simplicity, we describe three examples, illustrating different contexts in which kin recognition occurs.

3.3.4.1 Mutual recognition

Related individuals of the tunicate, *Botryllus schlosseri,* may settle near one another, and sometimes fuse. Recognition depends on sharing an allele at a highly polymorphic histocompatibility locus (Grosberg and Quinn 1986). Fusion of genetically similar colonial coelenterates probably depends on a similar mechanism. Note that in such cases there is no conflict of interest between the interacting individuals, and both provide information.

3.3.4.2 Nest-mate recognition in paper wasps

Like many social insects, paper wasps (*Polistes*) discriminate between colony mates and others (Gamboa *et al.* 1996). At eclosion, young wasps absorb hydrocarbons from their nest. These labels may be genetic (from a nest-mate) or environmental (e.g. from plant fibres) in origin. But, surprisingly, *Polistes* is capable of more than merely recognizing that another individual belongs to the same social group, using genetic or environmental cues. Queens and workers of *P. fuscatus* belong to colonies with a linear dominance hierarchy determining how food, work, and reproduction are divided. Tibbetts (2002) has shown that this depends on individual recognition of black and yellow facial and abdominal markings. Individuals whose markings were altered with paint were treated with greater aggression: this aggression declined with time as the changed marking became more familiar.

3.3.4.3 Altruistic behaviour and recognition of relatives

Altruistic behaviour, determined by Hamilton's rule ($rb > c$), does not require information about the relatedness of particular individuals: for example, altruism towards littermates could depend on average relatedness. However, a potential altruist can benefit from more precise information (e.g. if it can distinguish between half sibs and full sibs): in some cases, it is known that such information is used.

For example, Mateo (2002) has studied this type of information transfer in Belding's Ground Squirrel. This is a group-living species, in which females behave nepotistically (in cooperative territory defence and alarm call production) to close kin but not to distant kin: males disperse when adult, and show no altruistic behaviour (Sherman 1981). The information is transmitted by the secretions of two glands: it is not clear how genetic relatedness is encoded. Mateo collected secretions from these glands, singly or together, on polyethylene cubes, and presented them to subject animals. The more similar a subject animal was genetically to an unfamiliar donor, the less time it spent investigating the cube: Mateo used this fact to measure the squirrels' powers of discrimination. She found that both females and males were able to discriminate, for example, between siblings, cousins, and non-kin.

Thus, females were able to use the gland secretions to make the discrimination that they use behaviourally (between close relatives and others), and also a discrimination that they do not use (between distant kin and non-kin). Males were also able to make these discriminations, although they show no nepotistic behaviour. Mateo found a similar ability to discriminate in a related species, Golden-Mantled Ground Squirrel, which shows no nepotistic behaviour in either sex. It seems that the gland secretions are not a 'signal' of relatedness, in the sense of a structure that evolved because it conveys the information, but rather a cue that is present in both species, and is exploited in only one species: alternatively, Golden-Mantled Ground Squirrels may be revealing the ghost of nepotism past.

The absence of an evolved signal in such cases is perhaps not surprising. Although the potential altruist would benefit from more precise information about relatedness, the potential recipient of altruism would not necessarily do so. Thus, suppose that the relatedness estimated from a cue is r_0, and that a signal provides a more precise estimate, r_1. The potential recipient would benefit by signalling if $r_1 > r_0$, and lose if $r_1 < r_0$: that is, a potential recipient would gain by providing more precise information if it was indeed closely related, but not otherwise. But, in general, the recipient will not have that information.

3.4 Conclusions

One explanation, the handicap principle, for the reliability of signalling is that the signal is costly, either to produce or in its consequences. For this to work, it is necessary that a given signal is relatively more costly for low-quality individuals. Minimal-cost signals can also be reliable. This requires that signaller and receiver place the different possible outcomes of the interaction in the same rank order of preference. Even if they prefer different outcomes, an ESS may still be possible if liars are punished, if signaller and receiver have an overriding common interest, for example in avoiding an escalated contest, or if there are repeated interactions between the same individuals.

4

Indices of quality

4.1 Introduction

A crucial distinction we have tried to draw has been between 'indices', signals that are reliable because they cannot be faked, and 'handicaps', signals that are reliable because they are costly to produce. The distinction is clear enough in a model, and there are examples that fall easily into one or other category. Fights between funnel-web spiders are settled in favour of the heavier spider if the weight difference is greater than about 10 per cent. A spider can signal its weight cost-free by vibrating the web. Female stalk-eyed flies prefer males with long eye-stalks—eye-stalk length is associated with good condition, and hence with 'good genes', because it would be too expensive for low quality males to grow long eye-stalks. These are examples of an index, and a handicap, respectively. But not all real cases are so easily classified.

In this chapter, we first discuss the pitch of the roars of Red Deer as a classic example of an index, and point out some complications that arise even in this well-studied case. Section 3 discusses a problem with the evolution of indices. The problem is easily stated, but less easily answered: if an index gives accurate information about quality, why should a low-quality individual provide the information? The answer appears to be that an individual that does not signal is treated as one of low quality. In Section 4 we give examples of indices in a variety of contexts—indices of condition, of size and RHP, of 'performance', of parasitism, and of ownership, and compare the form of the signals used in contests and in mate choice. After this review, in Section 4.5 we return to the problem of distinguishing between indices and handicaps, and in Section 4.6 we discuss some problem cases—stotting, fluctuating asymmetry, and the display of weapons.

4.2 Are mammalian sounds reliable indices of size?

In Chapter 1, we used the pitch of the Red Deer's roar as a paradigm example of an index. This choice must now be justified; we think it can be done, but the story is not a simple one.

First, how are mammalian sounds produced? According to the 'source-filter' theory (Fant 1960), the sound is generated by the vibration of the vocal folds in the larynx. The fundamental frequency (F_0) generated depends on the length and mass of the vocal folds. The sound is then filtered through the vocal tract, extending from the larynx to the mouth. The air in the vocal tract has certain resonant frequencies, or 'formants', which determine the quality of the sound finally produced. By altering the form of the vocal tract, humans, and to a limited degree other primates, can modify the sound; in particular, vowel production by humans depends on this ability (Lieberman 1984).

The fundamental sound frequency, F_0, is a poor guide to body size (Lass and Brown 1978, for humans; McComb 1991, for Red Deer), because the size of the larynx and attached hyoid can vary independently of body size, as happens in male humans at puberty; as an extreme example, the Red Howler Monkey has a larynx and hyoid equal to the whole skull in size. However, Fitch (1997) argues that although F_0 is a poor guide to size, the formant frequencies, and in particular the spacing between formants, or 'formant dispersion', is a reliable one. In Rhesus Macaques, he found, as predicted by source-filter theory, that as the vocal tract lengthened the formant dispersal decreased, leading to a close negative correlation between formant dispersion and various measures of body size. This correlation exists because, unlike larynx size, the length of the vocal tract cannot easily be altered.

Things are a little more complicated when we turn to the roaring of Red Deer. In Red Deer, as in macaques, formant dispersion is a good guide to body size. When it roars, however, a stag lowers the larynx until it meets the sternum, thus reducing the formant dispersion (Fitch and Reby 2001). Since all stags do this, formant dispersion remains a good indicator of size. At one time, presumably, no stags lowered the larynx, and it was then also the case that formant dispersion was a reliable index. Then some stags evolved the ability to lower the larynx, thereby giving an unreliable but advantageous signal. However, selection would soon result in all stags doing it, and the signal would become reliable again. Reby and McComb (2003) analysed the roars of free-ranging red deer, and their relation to age, body weight and reproductive success. They found that the fundamental frequency, generated by the larynx, was a poor guide to body size, but that the minimum formant frequencies (at full retraction of the larynx) decreased with age and body weight, and that low formant frequencies were associated with high reproductive success. They conclude that '. . . acoustic cues . . . rendered honest by an anatomical constraint limiting downward movement of the larynx, provide receivers with accurate information'.

To summarize, the roar is today a reliable index, but during evolution a modifying feature evolved, exaggerating the signal. Since the feature is now fixed in the population, the roar is again a reliable index, because of a physical constraint on the length of the vocal tract—the larynx cannot be lowered further because of the sternum. However, what becomes of the claim that an index 'cannot be faked' if that is precisely what happened when larynx-lowering first evolved? In principle, all signals could be faked; taking two other examples mentioned in Chapter 1, perhaps one day funnel-web spiders will pick up a rock before vibrating a web, and tigers will stand on a box when scratching a tree. In a model, the phrase means no more than that it is *assumed* that the level of the signal cannot be faked. Applied to the real world, it means (roughly) 'there are no members of the present population that can fake the signal, and it is possible to suggest constraints that prevent them from doing so (e.g. spiders cannot carry rocks; tigers cannot find boxes; the larynx cannot pass the sternum)'.

The ability of Red Deer stags to lower their larynx when roaring appears to be one of two exaggerators of formant dispersal, since the resting position of the male larynx is already lower than the typical mammalian position. Stags share this character with male Fallow Deer and humans of both sexes. The claim that a lowered larynx is diagnostic of speech and language, which has dogged studies of our hominid ancestors, is thus incorrect. Size exaggeration provides an intriguing, non-linguistic explanation for the descent of the human larynx (Fitch and Reby 2001).

Red Deer roaring contests usually end with one stag retreating, but sometimes escalate into a parallel walk, in which both stags silently pace up and down side-by-side. Similar lateral displays are very common among ungulates and presumably provide an opportunity for assessing body size (and body condition, Jennings *et al.* 2002). In the context of exaggerators, it is interesting that stags often raise their hair during a parallel walk (Clutton-Brock and Albon 1979).

It is important to distinguish between an index, which increases the accuracy with which some quality of interest is perceived, and what we have called an 'exaggerator', which makes some quality appear greater than its true value. A feature which increases apparent size, such as raising the fur or lowering the larynx, is an exaggerator. Such a feature, once it arises, will rapidly be established in all members of a population. We should, therefore, be on the look out for features of an index, like the lowering of the larynx in Red Deer, that evolved in the past because they exaggerated the signal.

4.3 The evolution of indices

How did indices evolve? As Hasson *et al.* (1992) point out, there is a difficulty: although it will pay senders of high quality to signal, it will not pay senders of low quality. We are unable to offer a satisfactory model that meets this difficulty, essentially because it is hard to decide what assumptions to incorporate in such a model. In particular, could the giving of a signal be conditional on quality? That is, should the model assume that individuals of low quality have the option of not signalling? If so, would receivers interpret the absence of a signal as evidence of low quality? Then, once an index is established in a population, it may no longer be an option not to use it. A stag that did not roar would probably be treated as of low quality. In the mating context discussed in Section 4.4.3, a male *Drosophila subobscura* that did not dance would not mate. But how did such signals originate? One possible scenario is that an index can originate as an action or structure that exaggerates some feature that is already being used as a cue to some quality of interest to the receiver. For example, a cat exaggerates its apparent size by arching its back and fluffing its fur. It is easy to see why such exaggerating signals should spread. Once established in the population, however, they may act as indices, providing more accurate information. But these are speculations. It is clear that signals that are reliable because they cannot be faked occur in a number of contexts, but more work is needed to explain their origin and stability.

4.4 Indices in different contexts

4.4.1 Indices of condition

One example of such an index, the crop of the Bald Eagle indicating whether the bird has fed recently, was described in Section 2.1. We now give some further examples.

Male jumping spiders, *Plexippus paykulli*, display by exposing the ventral surface of the abdomen to rivals and potential mates (Fig. 4.1). In males, but not in females or juveniles, the abdomen has a dark central patch and pale margins. Taylor *et al.* (2000) measured the width of the central patch and of the pale margins in spiders collected from nature, and spiders starved for varying periods in the laboratory. They found that the width of the central patch varied little if at all with condition, whereas the pale margins were wide in well-fed spiders and shrivelled in starved ones. In the spiders from nature, the coefficient of variation of the width of the margins was 2.5 times that of the total abdomen width. The authors conclude that the patterning of the abdomen makes the pale margins more accurate indicators of condition, and suggest, reasonably, that the exposure of the abdomen is a signal of a male's condition, which may be of interest to a female as an indicator of heritable foraging ability or health. There remains the question why a male does not exaggerate the width of the pale margins by developing a narrower central patch; they suggest that this would signal small overall size, which could be disadvantageous. They did not record the response to the display of potential mates, and of rival males; it would be interesting if females are influenced by the pale patches (condition) and males by the width of the central patch (size).

Discussing these results, Taylor *et al.* argue that there is no reason to think that the signal is costly, and so no reason to regard it as a handicap; instead, the signal is reliable 'by design'. They refer to the body pattern as an 'amplifier', because it makes differences in condition easier to assess; we agree with their interpretation, but would call the signal an index. In their discussion, they refer to the description by Zahavi and Zahavi (1997) of the white patch on the rump of the Waterbuck, which is round in healthy animals but a pointed ellipse in emaciated ones. The patch, therefore, makes differences in condition more apparent. Taylor *et al.* argue that the patch should be interpreted as a cost-free signal of condition, analogous to the abdominal patches in *Plexippus*, and not as a handicap, the interpretation preferred by Zahavi and Zahavi. Again, we agree that the signal is best thought of as an index.

Roaring by Red Deer is not just an index of body size (Section 4.2). Harem-holding males often have to roar repeatedly at intruders by day and night: the energetic costs and reduced opportunity to forage means that males lose body condition and can become 'rutted out', lowering their roaring rate. Challengers rarely press home attacks against individuals that out-roar them (and rarely win if they do so), but often challenge

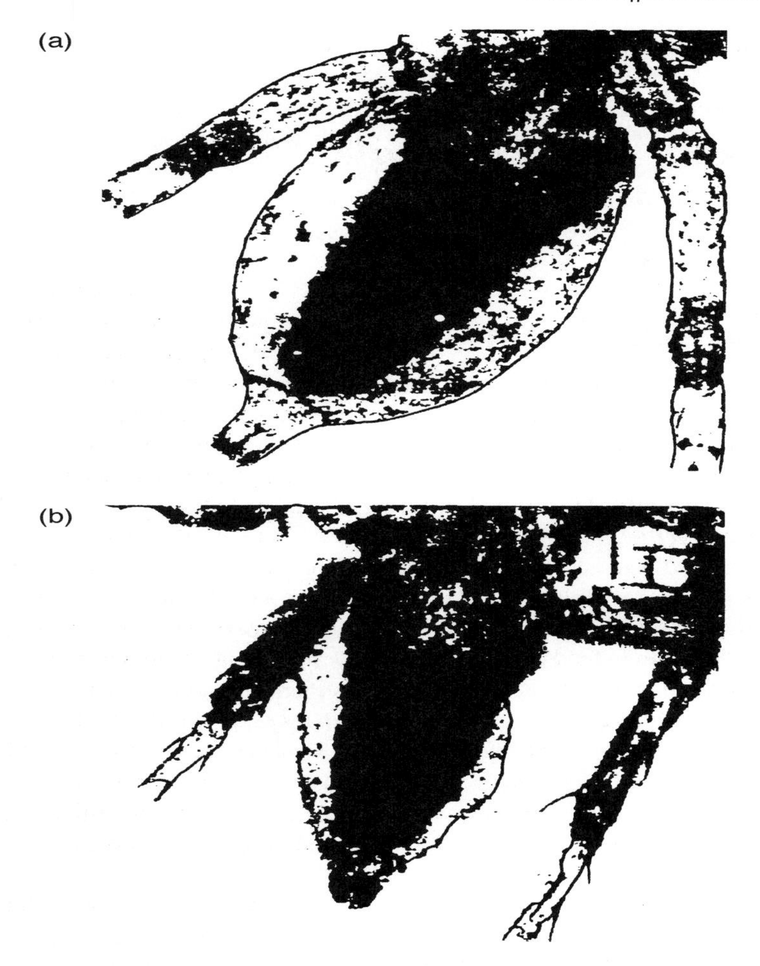

Fig. 4.1 Abdomen pattern as an index of condition in the salticid spider, *Plexippus paykulli*. Above, the day after feeding; below, after 20 days without food. Males display their abdomen pattern during courtship (after Taylor *et al.* 2000).

harem-holders that cannot match their roaring rate (Clutton-Brock and Albon 1979). For completeness, we note that males are not the only receivers of roars: females prefer to mate with males with high roar rates (McComb 1991) and can discriminate between the roars of current harem holders and other stags (Reby *et al.* 2001).

Parallel walks, as mentioned above, may give rival Red Deer stags another chance to assess each other's condition. Reserves of fat and muscle are often prominently positioned in mammals (Pond 1978). The Giant Deer, usually misnamed the 'Irish Elk', is famed for its colossal antlers (Clutton-Brock 1982). It was depicted by our

Neolithic ancestors as having a large—and apparently boldly coloured—hump situated over the shoulder blades (Lister 1994). This was far larger than the slightly humped back required by the spines on the anterior dorsal vertebrae (Dimery *et al.* 1985). Gould (1998) suggested that this hump 'would work especially well as an intimidating device', without giving reasons. We are tempted to be even bolder: the hump may have been an exaggerator of body size, or it may have been fat-filled and an index of body condition. Unfortunately, we are 10,000 years too late to test our hypotheses.

4.4.2 Indices of size and RHP

We referred in Chapter 1 to several examples of signals that act as indices of size—the depth of the croak in Common Toads, tigers scratching a tree as high as they can reach to mark their territory, and funnel-web spiders vibrating a web. The spider example is particularly clear because Riechert (1978) demonstrated that, if a weight is attached to the back of a spider, that spider will win contests it would otherwise lose, without physical contact.

Big-clawed Snapping Shrimps have a greatly enlarged claw which they use to produce an extremely fast water jet capable of stunning or killing small invertebrates (Versluis *et al.* 2000). Most contests consist of a pushing match, and are usually won by the larger individual: not surprisingly, most contests are settled before the contestants start shooting water at each other (Herberholz and Schmitz 1998). Shrimps assess each other's size using a display in which they spread their claw. Some males have claws that are large relative to their body size and likely fighting ability: such males perform the claw display more often. Thus, they win some disputes by display that they would lose if the contest continued to a pushing match, but they are also more likely to end up in more escalated interactions. Thus the claw display is used in assessment of competitive ability, but is not a perfectly honest signal (Hughes 2000).

In this chapter, we chose the pitch of the roar of Red Deer as our paradigm example of an unfakeable signal for two reasons: first, the quality of the sound produced does accurately predict size, for physical reasons that are well understood, and second, although today a stag cannot lie about its size, lying must have occurred in the past. When roaring, a stag lowers its larynx until it hits the sternum, thus lengthening the windpipe and making the stag sound larger. The first stags to do this were exaggerating their apparent size: now that all stags do it, the signal is again reliable. Such 'exaggerations' are likely to be a common feature of indices. In the present context, many animals have manes and dark lines down their backs, as well as movements such as arching the back, that make them appear larger. Summarizing, indices of size are widespread among animals, and often accompanied by exaggerating features.

Indices of size are also present in protracted contests of the kind that will be discussed in Chapter 6, consisting of a long exchange of signals of varying kinds. Following Enquist *et al.* (1990), we will argue that some of these signals provide information about relative RHP. As a contest continues, the participants acquire a more and more accurate estimate of their chance of winning an escalated fight: the contest ends when one of the contestants is convinced that its RHP is lower than that of its opponent. Some support for this interpretation is given by data on protracted fights between cichlid fishes over dominance.

In contests between groups, one might expect that RHP would be strongly influenced by the size of a group, and that the outcome of contests would be influenced by signals conveying information about numbers. This idea was confirmed by McComb *et al.* (1994) in a study of contests between groups of female lions, in playback experiments using recordings of single females roaring, and of three females roaring in chorus. They found that a single adult female was less likely to approach playbacks of a chorus of three lions than of a single lion, and that when they did approach they did so with greater caution. Defending females also took into account their own numbers, being more cautious if their own numbers were low.

Another context in which information about RHP may be exchanged is in the display of weapons. However, we are uncertain about the correct interpretation of such displays. We, therefore, postpone discussion of them to Section 4.6, which deals with some problem cases.

4.4.3 Performance indices

We have used the courtship dance of *D. subobscura* (Fig. 4.2) as an example of an index: females rejected males that lagged behind in the courtship dance, because they were aged, inbred, or carrying a mutant affecting locomotion or vision. Vigorous dancing or pursuit is a common feature of courtship, yet, although the *D. subobscura*

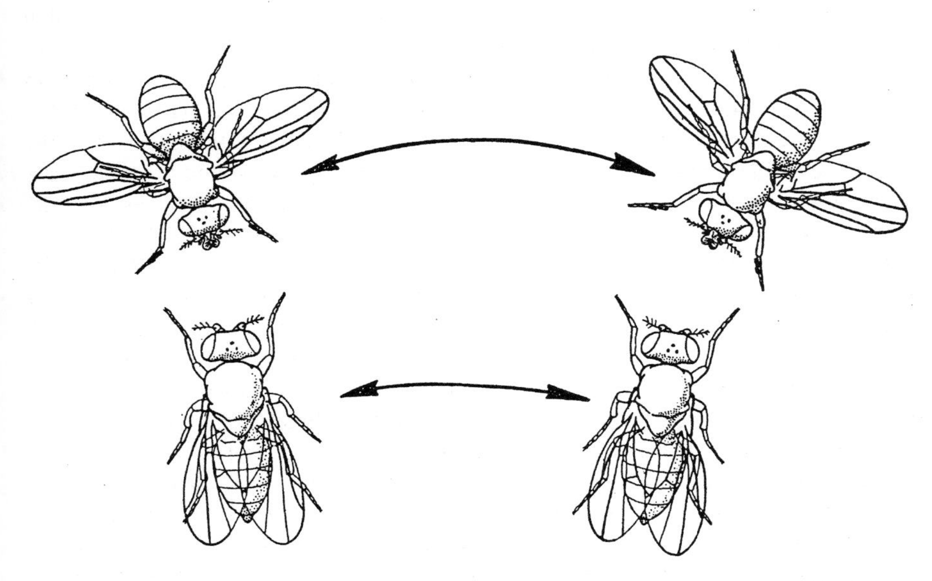

Fig. 4.2 Courtship dance of *Drosophila subobscura* (from Maynard Smith 1956).

report is almost 50 years old, there have been few subsequent reports (if any?) that females use male performance to select mates. This may be so because such performance tests are indeed rare. Or it may be that the possibility has seldom been investigated. Certainly, most ethologists reacted negatively to Maynard Smith's original report: in those days, whether or not an animal performed an act was determined by 'motivation', so it was argued that aged or inbred males had low motivation—an implausible explanation because an inbred male that had been courting unsuccessfully for half an hour was likely to attempt (unsuccessfully) to force a copulation.

What is needed is to show that a female's acceptance of a male does, or does not, depend on his performance. This is often difficult. For example, when a receptive female Silver-washed Fritillary butterfly encounters a male she releases a pheromone and then flies off in a straight line parallel to the woodland floor. The male follows, repeatedly swooping under her to emerge just in front of and above her head, before diving down again. He thus flies much further than she does. If the female accepts her suitor she simply drops to the ground to mate: otherwise she zig-zags away in flight. It would not be easy to demonstrate that the female's response varied with the male's performance.

Many courtship dances and pursuits are performed by both sexes. This could be because a female can most easily assess a male's performance by comparing it with her own. Or, if both parents invest heavily in the young, both sexes may have an interest in assessing their partner's fitness. Breathtaking examples of vigorous mutual courtship occur in the sky-dancing of birds of prey such as the Peregrine Falcon or Hen Harrier, in which both parents invest heavily in their offspring.

Some courtship dances do not rely on active female participation. When a male funnel-web spider walks onto a female's web, he usually starts to sway his abdomen from side to side and to flex the web with his legs. These actions can induce a cataleptic state among females. Successful males swayed their abdomens at higher frequencies than unsuccessful ones (Singer *et al.* 2000): thus, this is one of the rather rare cases in which a criterion determining female choice is known, although it is not clear what relevant quality is being indicated.

A quite different behaviour that might be explained as a performance index is courtship feeding, but there is little evidence to support this interpretation. Nuptial gifts in insects are reviewed in Section 5.4.2. Most cases can be explained either as a tactic by the male to prolong copulation and increase sperm transfer, or as sensory exploitation of the female. Curiously, the only case in which there is evidence that females use nuptial gifts to choose between males is in *Drosophila subobscura* (Steele 1986*a*,*b*). To sum up, there are many features of courtship that one is tempted to interpret as performance indices, but little evidence one way or the other. It is a topic that will repay investigation.

Performance indices may be relevant in contexts other than courtship. For example, stotting (see Section 4.6.1) appears to be an index of running ability, used by prey to deter pursuit. What of contest behaviour? In *The Casebook of Sherlock Holmes*, an aggressive visitor to Baker Street picks up the poker and bends it into a loop: Holmes responds calmly by bending it straight again, albeit only after the aggressor has left. Are analogous signals used in aggressive encounters between animals? Can one perhaps interpret the mouth-wrestling of cichlid fishes (see p. 102) in this way?

4.4.4 Parasites

The suggestion that females may use bright male ornamentation as an index of heritable parasite resistance originates with Hamilton and Zuk (1982). The idea is an attractive one. As the authors point out, the likelihood of a continuing evolutionary arms race between host and parasite would explain the maintenance of additive genetic variance for the selected trait—always a difficulty for 'good genes' models of sexual selection.

There is extensive evidence that parasite infection can be detrimental to male sexual ornaments (e.g. Merila *et al.* 1999), and that females mate preferentially with bright males (e.g. Petrie *et al.* 1991). Although required by the Hamilton–Zuk hypothesis, these observations could be explained in other ways—for example, if females were choosing bright males for reasons having nothing to do with parasites (e.g. because of Fisher's 'runaway' process), and the effect of parasites on brightness, although real, has nothing to do with the evolution of sexual ornaments. A further difficulty in establishing the hypothesis is that, even if bright plumage is used by females as an index of freedom from parasites, there are two kinds of reason, 'direct' and 'indirect', why females might choose bright males. The direct reasons are that, by choosing an uninfected male, the female avoids the risk of infection to herself and her offspring, and in some cases may ensure more effective paternal care. If this is what is happening, it is still true that bright plumage has evolved, at least in part, as an index of parasite resistance. But the full Hamilton–Zuk hypothesis also requires heritability of the resistance: there must be an indirect benefit to choice, in the transmission of 'good genes' for resistance to her offspring.

Barber *et al.* (2001) have provided convincing evidence of such an indirect effect in the Three-spined Stickleback. Using *in vitro* fertilization, they compared maternal half sibs, sired by a brightly coloured and a dull male, and raised without parental care. In this way, they eliminated the possible environmental effects, including the effect of egg size and quality, and parental transmission of parasites. They measured growth rate, and resistance to infection by a cestode parasite, *Schistocephalus solidus*, measured by percentage of brood infected. Offspring sired by brightly coloured males grew less rapidly than their half sibs, suggesting that the development of bright colouration is expensive. However, these same offspring were more resistant to infection, indicating that genes for bright colouration and parasite resistance are associated in the same males.

The explanation seems to be as follows. In a population subject to parasite infection, bright colouration will be associated with genes for resistance because males lacking such genes will be infected, and therefore, less brightly coloured. Hence, provided that there is additive heritability of resistance, it pays a female to mate with a brightly coloured male, because she will then have resistant offspring (of both sexes). As the authors point out, the different effects on parasite resistance and growth rate in uninfected animals provide a reason, additional to a possible host–parasite arms race, for the presence of additive heritability for bright colouration.

Additional evidence for the role of male ornamentation as an index of parasite resistance would be provided if it could be shown that the form of male ornamentation is peculiarly adapted to reveal infection. One possibility, mentioned by Hamilton and Zuk, is the existence of bare patches of skin, that may expose the colour of the blood, in otherwise furry or feathered animals. We know of no convincing evidence that such patches do, in fact, function in this way, but the suggestion is plausible, and evidence would be of real value. However, there is evidence that bare patches of skin can function in revealing parasite loads during courtship. Male Sage Grouse display at spectacular leks: although blood parasite loads do not correlate with mating success (Gibson 1990), males with many ectoparasitic lice gain few matings. Females assess mite load by detecting louse-induced haematomas on the male's oesophageal air sacs, and can be fooled by painting spots on the bare skin (Boyce 1990; Johnson and Boyce 1991). These paired olive-green air sacs are very obvious during courtship display: they are inflated so that they protrude from the pure white breast feathers, are used to make weird hooting sounds, and the leading edge of the wings are scraped across bristle-like feathers surrounding them. Thus, males act as if drawing attention to a cue used for female choice.

Throughout this discussion, we have referred to signals of the kind proposed as 'indices', despite the fact that some later authors have referred to them as 'handicaps'. In doing so, we are following the lead of the original authors, Hamilton and Zuk. Quoting Zahavi (1975), they remark that 'use of the word handicap . . . seems unfortunate', explaining that the 'female's object being not handicap but a demonstration of health that cannot be bluffed'—a perfect definition of what we later called an index.

4.4.5 Indices of ownership

The original 'Bourgeois' strategy, 'If owner, play Hawk; if intruder, play Dove' (Maynard Smith and Price 1973) was based on a model which assumed that all contests were between an owner and an intruder. If this was true, an index of ownership would be irrelevant, because an intruder already knows that its opponent is an owner. But the assumption often does not hold—for example, if the contested resource is a territory, both opponents are sometimes intruders. But contests over territories are almost always won by owners. Why should this be so?

Suppose that a territory cannot be shared, and that it is worth V (for the moment, V is the same for owner and intruder). If both contestants fight to the limit of their ability, the expected cost is C, where $C > V$; we return to this assumption later. A small fraction p of all contests are between two intruders. The following strategy is an ESS. If owner, fight up to a maximum cost of C; if intruder, fight up to a cost c, where $c \ll V$. The value of c will depend on p, and is likely to vary between individuals: it will not be zero because intruders want to win against other intruders. This strategy will be realized if individuals display for a length of time that increases with the time for which it has occupied the territory, starting with a low and variable c up to a maximum C. We will call this the 'Owner fights harder' strategy.

Several criticisms have been made of the original Bourgeois strategy. One is the rarity of the anti-Bourgeois strategy, 'If owner play Dove, if intruder play Hawk'. But if this strategy were adopted, as soon an intruder became an owner, it would be turned out by another intruder: no-one would hold a territory long enough to benefit from it. It is not surprising that there is only one reported example of this strategy (Burgess 1976). In the social spider, *Oecibus civitas*, individuals occupy holes in a communal web. If a homeless spider enters a hole, the occupant leaves at once. Presumably there are more holes than spiders.

One objection that has been made is that the value of the territory is greater to the owner. Some of the reasons suggested for this are unconvincing. For example, Dawkins and Krebs (1978) used the example of the nectivorous Hawaiian honeycreeper *Loxops virens* (Kamil 1978). Owners revisit flowers at long time intervals, but intruders often visited recently depleted flowers and as a result achieved only about two-thirds as much nectar per minute as the owner. The idea that this asymmetry should influence territorial contests is absurd: if the intruder expelled the owner it could forage systematically. But there are reasons why a territory may be more valuable to the owner: for example, the owner has paid the costs of settling boundaries with neighbours. Curiously, however, even if the territory is more valuable to the owner, this is quite irrelevant. Thus, suppose that it is true. Then we would expect to see the owner fights harder strategy. But we would see this strategy anyway, even if no such difference in value exists. Thus, a difference in value would not alter behaviour. This conclusion may seem contra-intuitive. But it is not reasonable to suppose that an owner thinks 'this territory is more valuable to me than to him, so I will fight harder'. Value could alter evolution only because it causes selection to affect owner and intruder differently, and so causes them to behave differently. But this does not matter, because at an ESS they would behave differently anyway, adopting the 'Owner fights harder' strategy.

To test the 'value' hypothesis, Tobias (1997) performed a series experiments on territorial European Robins, in which an owner was removed and replaced by a newcomer, and the owner was then released after varying periods of time. In winter, when both sexes defend individual territories, dominance shifted gradually from removed owners to newcomers, and contests lasted longest when the owner had been absent for 4–7 days. Removal of newcomers, followed by replacement by another individual, confirmed that the key factor was how long the newcomer had been in residence and not on how long the original owner had been held in captivity, The reversal in dominance coincided with the time taken for newcomers to settle territory boundaries and return to normal foraging rates. These experiments elegantly demonstrate that the time for which an individual will compete increases continuously up to a maximum. We do not agree, however, that a gradual rather than an abrupt reversal of dominance is a prediction unique to the value hypothesis (Krebs 1982; Tobias 1997). After all, the rule 'I own this territory because I have not been challenged in the 30 seconds since I arrived' is implausible. Imagine a newcomer encountering another Robin. It knows only how long it has been there itself, but cannot tell, in the absence of signalling, whether its rival is the old owner or another intruder, perhaps a neighbour. It is plausible that the best strategy is for the newcomer to make some minimal challenge when it has just arrived, but to behave more aggressively as time in residence increases.

It has also been objected that owners have a higher RHP, for example because they have won past contests. Provided that $C > V$, this objection is also irrelevant, for the same reason that a difference in value is irrelevant. Given a difference in RHP, owner and intruder would adopt the strategy 'Owner fights harder', but they would do that anyway.

But what if $V > C$: that is, the value of a territory is so high that it is better to fight to the limit than not to fight at all? Then, as Grafen (1987) pointed out, we would expect intruders to adopt the desperado strategy—fight to the limit. But if owners have a higher RHP, owners would still usually win, either because they won escalated fights, or because it no longer paid intruders to adopt the desperado strategy. This is a serious criticism. However, there remain many contests in which, almost certainly, $C > V$, yet the 'owner wins' rule holds. Many contests are over resources less valuable than a territory. In the case of territories, V will be relatively low if alternative, less valuable territories are available. For example, in Krebs' (1982) study of Great Tits, the most valuable territories were in woodland. If the owner of a woodland territory was removed, it was at once replaced by a bird which had previously occupied a less valuable territory outside the wood. In such cases, the value V of winning equals the difference in value between a good and poor territory. This is likely to be less than the cost of a fully escalated fight, and so the 'owner wins' rule is obeyed.

To summarize the argument so far,

1. The Bourgeois strategy, 'If owner, play Hawk; if intruder, play Dove', assumes that all contests are between an owner and an intruder: this will not be true of territories.
2. If some contests are between two intruders, it pays an intruder to compete at a small cost, c.
3. If the value of the territory is less than the cost C of an escalated fight, then the ESS, which we call 'Owner fights harder', is to display for a time that increases with the time an individual has occupied the territory from c to a maximum of C.
4. The 'anti-Bourgeois' strategy, 'If owner play Dove, if intruder, play Hawk', is not a plausible alternative, because no individual can hold a territory for long enough to benefit.
5. The fact that a territory may be more valuable to the owner is irrelevant.
6. If $V > C$, then it will pay intruders to adopt a desperado strategy, and it will only be true that owners usually win if, as is plausible, owners have on average a higher RHP. But there are probably many cases in which $V < C$: in such cases, owners will usually win whether or not they have a higher RHP.

Given that some contests are between two intruders, then a reliable signal of ownership would benefit both parties: it would save them the initial cost c. We describe two possible examples: scent-marking by mammals (Gosling 1982) and calls as conventional signals of ownership in wagtails (Davies 1981).

In mammals, olfactory signals are ideal ways of identifying both a territory and its owner. They can be long-lasting: a call or gesture finishes as soon as the signaller stops signalling, but an odour may hang about for days. Using scent, an owner can mark its territory boundaries and leave additional marks scattered around the rest of its territory. As long as its scent is individually distinctive, an intruder can identify the owner by the smell of its territory as soon as they meet (Gosling 1982; Gosling and Roberts 2001). Smells are therefore ideal indices of ownership. The point was nicely demonstrated by Gosling and McKay (1990). A male mouse was kept in a cage long enough to scent-mark it, and then removed. A second male was then placed in the cage for a short while, and the original male was then reintroduced. The second male at once submitted to the original one, recognizing it as the 'owner' by its scent.

Faeces and urine have the advantage as scent marks that they can be produced at minimal energetic cost. However, the amount of marking is constrained by the signaller's production of faeces and urine. An antelope, the Oribi *Ourebia ourebi*, demonstrates this constraint and provides an example in which males signal cooperatively (Brashares and Arcese 1999). Adult males defend territories, either alone or with male satellites. Both owners and satellites defecate on permanent dung middens along the territory borders or on top of the females' urine and faeces (which are scattered randomly). Owners with satellites benefit, since they mark with faeces less frequently than lone males, but deposit a greater volume per mark.

The argument that scent-marking allows intruders to identify owners can be applied to resources other than territories, even mobile ones. Some mammals mark their mates: for example, Maras of both sexes urinate on their partners (MacDonald 1985).

A scent mark will last longer if the key components are not too volatile: the problem here is that this will make the mark harder for a receiver to detect unless it is made conspicuous in some way. Many mammals leave their scent marks in prominent places. Otters leave their spraints (faeces) on objects that are conspicuous from otter eye level (e.g. a boulder or tree root by the water's edge). They concentrate their marking in areas which other otters are likely to visit such as the junction between two streams or where a canal towpath runs under a bridge (Chanin 1985). Male Roe Deer make a single clearing in their home range where they scent-mark. They use their antlers to fray the stems of bushes so badly that the plants wilt, and bite most of the plants' leaves off (Carranza and Mateos-Queseda 2001).

Scent-marking is a means by which a territory owner can make the territory like itself: an alternative way of indicating ownership is to make itself like the territory. This may explain why territorial mammals often roll on highly scented objects such as corpses and heterospecific scent marks, a behaviour as well known to many dog owners as its consequences are regretted.

If a mammal that rolls on a scented object can be interpreted as making a 'signal of ownership', what of vocal mimics? At least three explanations of vocal mimicry have been proposed:

1. The usual interpretation is that vocal mimicry is a means of acquiring a larger song repertoire (Catchpole and Slater 1995): they present evidence showing that a varied repertoire helps both in male competition and in attracting a mate.
2. The 'signal of ownership' explanation. For example, Australian Magpies mimic all vocal species (including humans) that are regular on their territories, but not temporary visitors (Rogers and Kaplan 1998). Since they can mimic calls extremely well after hearing them just once, this does not seem to be because unusual sounds cannot be learned: rather it is consistent with the idea that only regularly heard sounds are worth mimicking. But not all details of vocal mimicry are consistent with the idea that owners are trying to 'sound like their territory'. One of the greatest vocal mimics over much of central and Eastern Africa is the White-browed Robin Chat. In this case, the mimicked vocalizations are often used at inappropriate times, for example, nightjar calls at noon or the songs of migrants while their models are on their Eurasian breeding grounds (van Someren 1956). Even more dramatically, the large vocal repertoire of Marsh Warblers consists largely of mimicry, and involves songs from both their European breeding grounds and African wintering area: all these are sung at both sites (Dowsett-Lemaire 1979). These observations fit better with the 'acquiring a large repertoire' hypothesis.
3. Owners benefit if intruders treat the sounds of mimicked species as coming from the territory owner. This is consistent with the frequent assertion that mimicry is associated with living in dense vegetation (e.g. Marshall 1950).

To summarize, we accept that mimicry often functions to enlarge the mimic's vocal repertoire. But we are intrigued by the idea, suggested by observations on Australian Magpies, that mimicked sounds may function as signals of ownership.

We turn now to a case in which a unique call is given by territory holders, and respected by intruders: the behaviour has been well studied, but we are unable to give a fully satisfactory explanation of its stability. Male White Wagtails frequently defend winter territories wherever food is both dependable and localized (Zahavi 1971). The owners, which are usually adult, sometimes form a temporary association with a satellite (first winter bird of either sex or an adult female). Individuals which are neither owners nor satellites form flocks, which are frequently visited by territorial birds. This social system has been intensively studied in the Pied Wagtail (the British subspecies; Davies 1981; Davies and Houston 1981, 1983) in the particular case of males defending contiguous territories along the banks of the River Thames. An unusual feature of this study was the food supply: dead insects washed onto the mud at the water's edge. A wagtail could temporarily remove all the food from a stretch of mud, but food availability would gradually increase again to an asymptote. The key point is that immediately after a wagtail had visited a piece of mud, this was an unprofitable place to forage. Owners therefore walked a regular circuit around their territories, which usually included both banks of the river, taking about 40 min to revisit the same piece of mud. The best place to feed was immediately ahead of the owner where intrusion was readily detected: an intruder foraging further from the owner was safer but found little food. Thus, intruders would benefit from detecting the presence of an owner at low cost. Davies (1981) suggested that was why intruders called a loud 'chisick' (the species' usual call) and why they usually retreated if the owner replied 'chee-wee' (a call given only by owners). When feeding conditions were good—thanks to high insect abundance and suitable wind conditions—some intruders persisted and performed complex 'appeasement displays' together with high-pitched calls (see Zahavi 1971). A few of these individuals were accepted temporarily as satellites (see above). The owner and satellite walked around the territory about half a circuit apart so that a piece of mud was visited about twice as often as when the owner was alone. Although satellites sometimes chased intruders, the main assistance they gave to owners appeared to be by reducing the profitability of the territory to intruders. When food availability fell too low, the owner promptly evicted the satellite.

In this case, then, we have a clear signal of ownership, never made by non-owners. The advantage gained by an intruder in not using this call is, presumably, that it may gain by being accepted as a satellite, and does not provoke the owner into immediate aggression. But it is less clear what the owner gains by accepting a satellite, which it can apparently evict when it wishes. We need to know more about the signals exchanged in this and other species where owners share their territory (e.g. Ruff; Cramp 1983).

4.4.6 Signals in contests, and in mate choice

It is interesting to contrast the kinds of signals made in contests, and in mate choice. As a gross oversimplification, it could be argued that contests are usually settled by indices, and mate choice by handicaps. There are plenty of exceptions to this generalization, but it is true often enough to be worth discussing.

First, consider the signals used to settle contests. Such signals are often indices of RHP—in particular, of size or condition. Displays of weapons are also quite common. Such signals can often be interpreted as ritualizations of cues used to predict the intentions or fighting ability of an opponent. It is harder to point to handicaps used in contests. An exception is the use of 'risky' signals: a signal exposing a contestant to the danger of retaliation may be too costly, except for an individual of high RHP or motivation. We suggested earlier that musth in elephants is best interpreted as a handicap, but it seems to be unique. We can think of no example, in contest behaviour, of a structure, analogous to the tail of a Peacock or a Lyrebird or the eye-stalks of a stalk-eyed fly, that is used to settle contests but that is purely symbolic, in the sense that its form is unrelated to actual fighting ability. There are symbols used in contests—for example, the evanescent stripes of cichlid fishes (see p. 104), but they are used in the early stages of a contest.

In contrast, symbols seem to be common in mate choice. This is to be expected if the evolutionary mechanism is that suggested by Fisher (1930): all that is required is an initial, arbitrary preference by some females, later exploited by males. If choice is for 'good genes', indices may evolve. Signals indicating freedom from parasites are a clear example. Indicators of athletic or foraging ability are other possibilities.

The dance of *D. subobscura* is an example of the former, but it is hard to point to other cases. Courtship feeding seems not to be used as an index of foraging ability, at least in most cases: rather, from the male's point of view, it is an exploitation of female appetite, or a means of transferring nutrients to one's own offspring.

4.5 Indices and handicaps

In a formal model, the distinction between an index and a handicap is simple. An index is not costly to produce, and is reliable because, for physical reasons, it cannot be faked: a handicap is reliable because it is costly to produce, excessively so for low quality individuals. In real cases, however, the distinction is not always so clear: the difficulties are best illustrated by some examples.

Consider the courtship dance of *D. subobscura* (Maynard Smith 1956). If a male approaches a virgin female head to head, she steps rapidly from side to side, and the male attempts to keep facing her. If he succeeds, she will usually stand still, allowing the male to circle round and mount. If he lags behind (in the experiments, this happened because the male was inbred, aged, or had deficient eyesight), the female 'decamps'. The obvious interpretation is as follows. The dance enables a female to choose between males. She has been selected to choose a male with 'good genes'—that is, genes conferring high fitness. (Alternative theories of mate choice are discussed in Chapter 5, but the 'good genes' theory is probably appropriate in this case). There is no way in which she can choose on the basis of all the male's genes, or indeed on the basis of genes alone, ignoring environmental differences, so the best she can do is to choose on the basis of 'athletic ability', which is a component of total fitness. The dancing ability of a male is a measure of athletic ability, and so an index of good genes.

If this explanation is correct, males are being sexually selected for high athletic ability, of which dancing ability is an index. If there is a trade-off between athletic ability and other components of fitness, for example, fertility, the long-term result may be that males of high overall fitness will make a relatively higher investment in athletic ability, particularly in those features that make for dancing ability, than do males of low fitness. In principle, this could be true of any of the 'performance indicators' discussed above. If so, such indicators could reasonably be interpreted as handicaps. We, therefore, have a choice:

1. We can say that dancing ability remains a reliable index of one component (athletic ability) of a quality (good genes) of interest to the receiver, but, because only one component is being signalled, that component of the phenotype may be exaggerated relative to others, perhaps in a condition-dependent manner.
2. We can interpret the condition-dependent expenditure on athletic ability as a handicap, acting as a reliable signal of overall fitness.

We prefer the first alternative for two reasons. First, there is no evidence for condition-dependent expenditure on dancing ability or other performance indicators.

More important, the crucial distinction we are making between an index and a handicap concerns the reasons for their reliability: an index is reliable because it cannot be faked, and a handicap because it is costly. Dancing is a reliable indicator because it requires athletic ability (which is a component of overall fitness, the quality of interest to the female), and not because it is costly—it is not. A second example may help. We argued above that the pitch of the Red Deer's roar is best interpreted as a reliable index of size. But what the recipient of the signal is interested in is not size itself but RHP (resource-holding potential—in effect, fighting ability; Parker 1974). If size, rather than weapons, etc., is signalled, one might expect selection to exaggerate size at the expense of other components of RHP. But even if this happened, it would be clearer to describe roaring as an index of size, while accepting that size is only one component of RHP, essentially because the reliability of roaring depends on the physical constraints that prevent faking, and not on its cost.

To summarize, we suggest that the classification of a signal as an index or handicap should depend on the nature of the signal itself, and on whether its reliability is maintained by physical constraints or by costs. But we should also remember that if only one aspect of some quality of interest is signalled, that can lead to an alteration in the relative investment in different aspects of the phenotype.

4.6 Some problem cases

4.6.1 Stotting

To 'stot' (Fig. 4.3) is to leap in the air with all four legs held stiff and straight (Walther 1969). It is performed by many antelopes and deer, usually in response to the presence of a predator. It has been studied in the wild in Thomson's Gazelles by Caro (1986) and by FitzGibbon and Fanshawe (1988). How is it to be explained?

First, is it a signal at all? Could it simply be the best way to travel fast, or to change direction suddenly? The possibility has to be considered, because Mule Deer can reach similar speeds whether stotting or galloping, leading Lingle (1993) to argue that stotting enabled the deer to use obstacles to their advantage when evading predators. However, this cannot be the explanation in Thomson's Gazelles: FitzGibbon and Fanshawe observed that stotting ceased if the pursuing predators were closer than approximately 40 m.

If, as seems to be the case, it is indeed a signal, one possible explanation is that it tells a potential predator 'I have seen you, so you will only waste time if you continue to stalk me'. Such 'pursuit deterrance' signals certainly exist; see Hasson (1991) for a review. For example, a potential prey individual may look directly at a predator with ears pricked, and the direction of gaze may be emphasized by facial markings (Zahavi and Zahavi 1997). However, this too seems unlikely to be the explanation of stotting. Caro (1986) observed that gazelles do sometimes stot in response to a Cheetah, but found no evidence that the signal either provoked or deterred pursuit,

Fig. 4.3 Springbok stotting (from MacDonald 1984, vol. 2).

and FitzGibbon and Fanshawe (1988) observed that stotting was performed more frequently to African Wild Dogs, a coursing predator, than to Cheetahs, a stalking predator.

The natural explanation is that stotting is an index of condition and of escape capability: it pays the predator to pursue individuals of lower capability, and the prey to avoid potentially dangerous pursuit by signalling high capability. FitzGibbon and Fanshawe support this interpretation, and offer the following evidence in its favour. Several lines of evidence suggest that stotting ability is indeed correlated with ability to escape. Of the individual gazelles selected by Wild Dogs as the targets of pursuit, those that ultimately escaped were more likely to have stotted, and for longer durations, than those that were ultimately killed. Also, the dogs were more likely to succeed in killing a gazelle from groups showing a low frequency of stotting, suggesting that such groups contained a higher proportion of vulnerable individuals. It is harder to demonstrate in the field that low stotting rate is associated with poor body condition, but some support is provided by the fact that stotting rates were lower in the dry season when physical condition was poor.

It is also clear that the dogs were selecting as targets those gazelles with lower intensities of stotting. If two gazelles were running along side one another at approximately the same speed, the dogs were more likely to select the one with the lower stotting rate. More generally, the gazelles selected as targets stotted at lower rates than those in the group that were not selected. Occasionally, the dogs switched from one target gazelle to another: in four of five cases, they switched to a gazelle with a lower stotting rate.

If FitzGibbon and Fanshawes' interpretation is correct, then stotting in Thomson's Gazelle is an index of ability to escape a coursing predator. It pays the predator by enabling it to select a relatively vulnerable target, and it pays a high-quality signaller by enabling it to avoid a protracted and dangerous pursuit. As is usually the case with an index, the signal confers no benefit on low-quality signallers, but that is something they can do nothing about. The signal is a reliable one, but it is hard to see how it could be a handicap, unless it is interpreted as a signal to conspecifics, saying 'see what a splendid gazelle I am'.

4.6.2 Fluctuating asymmetry

Fluctuating asymmetry (FA) refers to non-directional differences between the two sides of almost-symmetrical structures. There has been much discussion recently about the possible role of FA in sexual selection. The data are hard to assess. We discuss the topic here because it illustrates some of the difficulties that arise in deciding whether a signal exists, and if so whether it is an index or a handicap.

It was suggested (originally by Mather 1953) on *a priori* grounds that FA is a measure of developmental homeostasis, or 'canalization' (Waddington 1957). Thus, a difference between the two sides reflects the failure of a given genotype to give rise to a constant phenotype in a constant environment. After lying fallow for over 30 years, this idea has recently been resurrected in the context of animal signalling. Møller (1992) reported that female swallows prefer males with more symmetrical tails, and suggested (Møller 1993) that there may be a more general tendency for females to select males of low FA, and hence of high general fitness.

This raises three questions:

Is there indeed a correlation between low FA and high general fitness?

The idea is plausible, but hard to test empirically, because of the difficulty of measuring fitness. We shall not discuss it further.

Is FA heritable?

Mohler and Thornhill (1997) claim, on the basis of a meta-analysis of the published data, that it is. However, their paper is followed by six commentaries, all to a greater or lesser degree critical of the way in which the data have been collected and interpreted. Perhaps the most serious criticism (Whitlock and Fowler 1997) is that a number of the studies rely on correlations between siblings: such correlations can be confounded by environmental and maternal effects, and in any case can at best provide evidence only of genetic differences in FA, but not of additive heritability (that is, resemblance between parents and offspring).

Two recent studies (Tomkins and Simmons 1999 on European Earwigs; Bjorksten *et al.* 2000, on stalk-eyed flies, *Cyrtodiopsis dalmanni*) were set up specifically to measure heritability on a male trait (forceps length in earwigs, eye-stalk length in *Cyrtodiopsis*) known to be sexually selected by females. The two papers reach remarkably similar conclusions. There is no evidence of additive heritability of FA for either trait, but clear evidence for additive heritablity of the size of the structure itself.

Curiously enough, however, there is good evidence in one context for additive heritability for FA, albeit very small. The most sensitive and reliable method of measuring the additive heritability of a trait is to measure its response to directional artificial selection. Selection experiments for FA in *Drosophila*, on bristle number rather than a continuously varying trait, (Mather 1953; Reeve 1960) resulted in a positive response to selection. Estimates of heritability, however, were low ($h^2 < 0.1$).

Do females prefer to mate with males with low FA?

There are two related difficulties in answering this question. First, it is hard to believe that females can detect some of the small asymmetries concerned, for example in sterno-pleural bristle number in *Drosophila*. Second, a correlation between low FA and mating success would arise even if females are unable directly to perceive or respond to differences in FA, if (a) females are selecting males of high fitness, using a cue or signal other than FA and (b) if there is a correlation, genetic or environmental, between the trait selected and low FA. For example, female *D. subobscura* reject inbred males because they cannot dance: if inbreeding increases FA, then there would be a correlation between mating success and low FA, but not because low FA influenced female choice. This objection can be overcome if FA can be experimentally manipulated, but that is often difficult to do.

The clearest evidence for female perception of, and preference for, low FA is the paper by Møller (1992), mentioned above, on swallows. He manipulated both the mean length, and the asymmetry, of the outer tail feathers of male swallows, by cutting and gluing, and measured the delay before pairing. Males with longer tails, and with less asymmetrical tails, paired earlier. The mean asymmetry before manipulation was 3 mm. In the experimental group with increased asymmetry, the difference was 20 mm—much larger than any differences occurring naturally. However, he also created a group with reduced asymmetry—0 mm instead of an average of 3 mm. These males paired earlier, indicating that females were able to discriminate naturally occurring differences in asymmetry.

Unfortunately, it is not clear that more recent work supports Møller's conclusions. On the basis of a meta-analysis of 65 studies and 42 species, Møller and Thornhill (1998) concluded that there is a 'moderate' relationship between low FA and mating success or attractiveness of males. However, most of the studies are open to the objection that the correlation between low FA and mating success may be caused by female preference for some cue other than low FA. Breuker and Brakefield (2002) measured the effects on female choice of the dorsal eye spots in the butterfly *Bicyclus anynana*, after experimental damage to the pupae with a fine needle had caused large differences in size and asymmetry. They found that large size, but not low asymmetry, favourably influenced female choice, although the experimentally caused differences were large compared to those occurring naturally.

One context in which FA has been experimentally manipulated is in the study of human preferences for symmetrical or asymmetrical faces. Several studies have manipulated the asymmetry of photographs of human faces. These have accidentally demonstrated just how difficult such experiments are. Early investigators confounded manipulations of asymmetry with facial 'averageness' and mean trait size. Swaddle and Cuthill (1995), therefore, manipulated asymmetry within a face without altering the mean size of facial features. Contrary to earlier claims, faces that were made more symmetrical were rated as being less attractive. This result, in turn, may be an artefact of unnatural feature shapes and skin textures introduced by image-processing. Certainly when Perrett *et al.* (1999) tried to avoid such problems, they found that increasing the symmetry of face shape increased the perceived attractiveness of both male and female faces. Strikingly, this pattern is still observed when the symmetry cues are removed by cutting the photograph in half (down the vertical midline, bisecting the nose, Scheib *et al.* 1999): this suggests that faces are also assessed as attractive on the basis of features which correlate with symmetry. Recent experiments using computer graphic faces confirm this result without the artificiality of viewing 'half-faces', but have not identified the non-symmetrical characteristics involved (Penton-Voak *et al.* 2001). The complexity of this field is highlighted by the marked inconsistencies about which faces observers find attractive. For example, women change their perceptions of male faces at different stages of their menstrual cycle and in different contexts—for example, presence or absence of partner (Penton-Voak *et al.* 1999). Turning to differences between observers, women who regard themselves as attractive have a stronger preference for symmetric male faces (Little *et al.* 2001). This behaviour may well be analogous to the condition-dependent mating strategies found in other species (e.g. Bakker *et al.* 1999). But its relevance here is to underline the need for future studies of facial attractiveness and FA to consider the interactions between multiple properties of the assessors and the faces that they view (Penton-Voak *et al.* 2001).

We do not, at present, find it possible to decide how often FA is influencing mating success, or, if it is, whether it is being signalled. However, suppose it proves possible, in some species, to show that FA is correlated with fitness, and that females perceive differences in FA and select males accordingly, would we regard this as an index or a handicap? Probably, neither: unless some display existed, whereby males made differences of FA easier to perceive, FA would be a cue, not a signal. Suppose, however, that it is confirmed that male swallows display their tails, and that females prefer symmetrical tails. This, presumably, would be an index of high quality, not a handicap. It would be a handicap only if females preferred males with highly asymmetrical tails: only a fit male could fly with such a tail.

4.6.3 Displays of weapons

Weapons are frequently displayed during contests: for example, wolves display their canines, Stag Beetles their 'antlers', and crabs their claws (Fig. 4.4). But first, what of the evolution of weapons as fighting organs? Some animals routinely fight viciously, using weapons capable if inflicting severe injury or death. Male *Idames* fig wasps fight for control of females emerging from a fig, and have enormous mandibles which can cut rivals in half (Hamilton 1979). Male Narwhals—the so-called Unicorns of the Sea on account of their spiralled tusk which is up to 3 m long and propelled by over 1.5 tonnes of whale—suffer terrifying injuries (Silverman and Dunbar 1980). We often regret never seeing a Unicorn: perhaps it is as well that myth portrays them as peaceful creatures. Male Gladiator Frogs have scimitar-like blades on their thumbs which they use to slash at their opponent's ears and eyes. Most males are battle-scarred and fights can be lethal (Kluge 1981). The case is a puzzling one. The contested resource is a small pool, built by the male, in which females lay their eggs. Hence, contests are presumably between the 'owner' of a resource and an intruder. One would therefore expect the willingness to escalate, and hence the outcome of contests, to be strongly influenced by ownership, but that escalated contests would occur when the intruder is larger than the owner. It would be interesting to know whether ownership, in fact, influences behaviour.

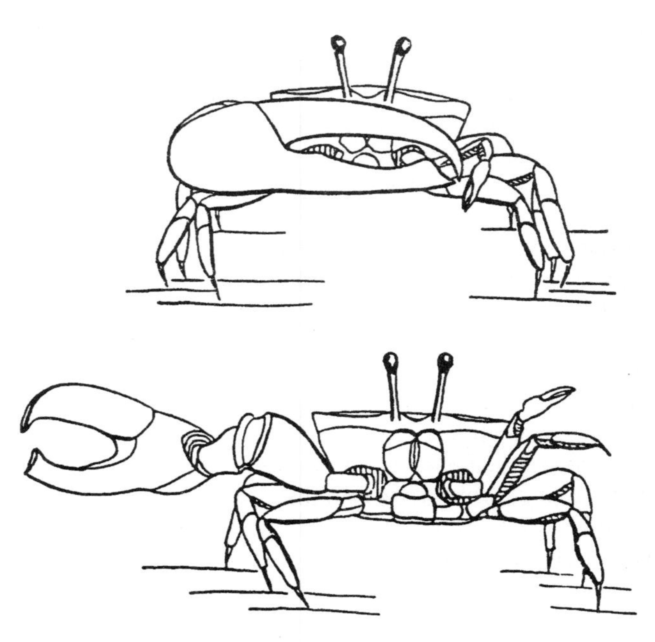

Fig. 4.4 Claw-waving display of a fiddler crab *Uca lacatea*, from Huntingford and Turner 1987.

Despite the existence of such lethal weapons, the early ethologists were struck by observations suggesting that contests were often harmless, ritualized trials of strength (e.g. Lorenz 1966). This conclusion was largely accepted by Maynard Smith and Price (1973), who argued that their ritualized nature could be explained if escalated aggression was met by retaliation. Although the notion of retaliation is sound (it is called 'tit-for-tat' nowadays), the account of ritualized fighting behaviour that gave rise to it was needlessly naive.

A more realistic account of the evolution of animal weapons—in particular, in ungulates—was given by Geist (1966). In larger ungulates, weapons used primarily for stabbing are atypical. Fights tend to be of one of two kinds: ramming matches (e.g. Bighorn Sheep, Musk Ox), or wrestling and pushing matches (e.g. most deer, many antelopes, elephants). In species in which a pushing match is the typical form of fighting, a contestant will take any opportunity that arises to attack the soft flank or belly of its opponent: for example, if a Red Deer stag slips over, his rival will immediately try to gore him (Clutton-Brock and Albon 1979). Geist suggests that the branching structure of antlers, and the twisted and ridged horns of antelopes, evolved as a defence against such attacks, by binding an opponent's antlers or horns.

What of the display of weapons? Geist (1966) discusses the display of horns in Bighorn Sheep. The horns, used in a ramming contest, may amount to 10% of the body mass. They are displayed by a male moving through a band in a stretched posture with his head extended, revealing the curls of the horns. Geist presents several lines of evidence showing that information concerning horn size influenced future behaviour. Perhaps his most convincing observation is of rams integrating into a new band, in which the fighting ability of other members is not directly known. In five such cases, the new ram interacted extensively only with other males whose horn size was similar to its own. It seems clear that the 'stretch display' does convey information about horn size, and thereby influences the behaviour of others: it is a signal. If, as seems likely, horn size is a reliable cue to fighting ability, then the stretch display acts as an index of RHP.

A simpler example of a display of a weapon that acts as an index of RHP was described by Sneddon *et al.* (1997). Male Shore Crabs with larger claws are more likely to win fights: such well-armed individuals display their claws more frequently during contests and often win without having to escalate to combat.

However, not all examples of an animal using a weapon as a component of a signal are easy to interpret. Elsewhere, we describe two cases in which such displays occur, but are to a greater or lesser degree dishonest. Male Snapping Shrimps display their enlarged claw in contests, but claw size is not an accurate index of body size (p. 50). In the fiddler crab *Uca annupiles* males that lose a claw replace it with a less effective claw of lower mass, which appears not to be detected by other crabs (p. 88). These examples cast some doubt on the idea that display of a weapon is an unfakeable index of RHP.

5

The evolution of signal form

So far, we have had rather little to say about the form of signals. In Chapter 2, we were concerned primarily with the cost of producing a signal, and the relation between the cost and reliability of the signal, but its form was ignored. In contrast, the form of an index, discussed in Chapter 3, is crucial, since an index only works if its intensity is causally tied to the quality being signalled. But many signals are not indices; we now turn to various ideas that have been proposed to explain their form.

5.1 Ritualization

Most signals probably evolved by the ritualization of cues that other individuals were already using to gain information (Chapter 1). The idea of ritualization has two aspects. First, the origin of a signal in a pre-existing cue, and second, the fact that the signal is often more stereotyped and repetitive than the original cue: we discuss these two aspects in turn.

Morris (1956) divided the cues that one animal might provide to another into those resulting from changes in physiological state and those that are behavioural. There are a variety of physiological responses that appear to have provided cues that have been ritualized into signals. In all the following examples the hypothetical cue was under the control of the autonomic nervous system:

1. *Thermoregulation.* Birds and mammals raise and lower their feathers or hair to help control their body temperatures. Usually their plumage or pelage is slightly raised away from the body, trapping an insulating layer of air. If the feathers or hair are erected, however, this layer is lost and heat is dissipated. Social interactions must often involve an increase in body temperature as a result of increased activity. There is thus scope for changes in feather and hair posture to provide cues suitable for ritualization into signals. Although many displays do indeed involve dramatic changes in feather or hair erection (e.g. Andrew 1972), it remains unclear how many evolved in this way. In particular, by erecting its feathers or hair in a display an animal makes itself appear larger. Thus, this component of the display may be an 'exaggerator' (as discussed in Section 4.1) of body size. If so, receivers would have been trying to evaluate body size rather than using thermoregulatory changes as a cue to increased body temperature.

2. *Respiration.* The increased activity likely in many social interactions may prompt an increased respiration rate. Rapid breathing might be loud enough to provide a cue that can be ritualized into a call. Similarly, the associated movements may provide the raw material for signals. For example, many fish displays highlight the gill covers which are often coloured or decorated with protuberances.

3. *Urination and defecation.* Extremely frightened mammals often relieve themselves. The use of urine and faeces to mark territory boundaries might have evolved from this because territory owners relieved themselves with fear at the edge of their range but were willing to fight intruders (Lorenz 1970). On the other hand, this is not the only possible explanation. Mammals have to relieve themselves at intervals, whether frightened or not. If potential intruders began to use urine and faeces as a cue that an area was occupied, it might pay owners to relieve themselves at territory boundaries.

4. *Pupil dilation.* The diameter of the pupils in the eyes does not vary only with light levels. For example, an African Grey Parrot constricted her pupils when saying a word that she had learned and when she heard the same word spoken (Gregory and Hopkins 1974). Humans are unconsciously aware of this cue in other individuals, assessing dilation of the pupils as a friendly gesture (Hess 1965). Women have in the past exploited this fact by adding drops of belladonna (an antimuscarinic drug) to their eyes to dilate their pupils. A side-effect of belladonna is that it inhibits the lens muscles of the eye and so increased attractiveness to men may have been accompanied by reduced discrimination of their suitors. The size of the pupil is more obvious if the surrounding iris is pale. Iris colour often varies between birds of the same species, but of different age or sex. For example, female Galahs (an Australian cockatoo) have pink irises against which their pupil diameter is easy to assess. Males, however, have very dark irises, almost as if they are trying to conceal their pupil diameter (Rogers and Kaplan 1998): the meaning of these signals is unknown.

5. *Yawning.* This is a puzzling case. In primates, the yawn is a signal of aggression. This makes good sense—the yawn bares the canine teeth, and may well have originated by ritualization. But what of human yawns? They are certainly not signals of aggression. There is evidence that yawning is unrelated to prior amount of sleep, but that it acts to maintain or increase arousal when the environment provides little stimulation (Baenninger 1997). This suggests that, as our ancestors ceased to use their teeth to fight with, the yawn ceased to be a signal that had an obvious, iconic meaning (although it is not clear why it could not continue to function as a symbol of aggression), and which was, therefore, free to acquire a physiological role. However, the human yawn may still have some role as a signal: like laughter, it appears to be infectious.

The range of behavioural cues that seem to have been ritualized into signals is as diverse as that of physiological cues. For example:

1. *Intention movements.* Before birds take off, they often crouch down on flexed legs, withdraw their head and raise their spread tail. The legs are then straightened, launching the bird into the air as it opens its wings. If a human or a predator slowly

approaches a bird, it may make several 'false starts' before fleeing. The movements could, therefore, act as a cue that the individual is likely to take off. Many species make such intention movements before leaving a flock without being disturbed; if they do so their companions usually ignore their departure, but if they do not the flock often flees in panic.

A wide variety of displays seem to be derived from these flight intention movements (Andrew 1956). Sky-pointing by gannets, boobies and some of their relatives (Fig. 5.1) is a well-studied example (van Tets 1965; Kennedy *et al.* 1996). In the Atlantic Gannet, the display precedes take-off from the nest site: the bird stretches its neck, pointing its bill vertically, while slightly moving its wings away from the body and lifting its feet alternately from the ground. Among boobies, however, sky-pointing occurs during courtship and is more complicated. The most extreme version is that of the Blue-footed Booby: the wings are opened and rotated at the shoulder so that their upper surface points forward and the tail is cocked until it also leans forward. The posture is so distorted that the outstanding ethologist Michael Cullen initially thought that it was mythical (Nelson 1978). As the displaying bird moves away, it raises its feet in a highly exaggerated manner, flaunting their bright blue webs. Thus, as Krebs (1987) said of ritualization in general, 'an ancestral dither may have evolved into a dance akin to that of a whirling dervish'.

2. *Protective movements.* When approaching each other primates often flatten their ears, retract their scalp and partially close their eyes. These behaviours, which tend to be most marked in subordinates, seem best interpreted as protecting vulnerable

Fig. 5.1 Ritualized flight intention movements in the Blue-footed Booby (from Cramp and Perrins 1994).

parts of the head. In some species these protective movements seem to have been ritualized into signals. For example, in many Ceboid and Cercopithecoid monkeys scalp retraction exposes brightly coloured skin under the eyebrow ridge (Andrew 1963).

3. *'Displacement' behaviour.* Animals often interrupt one behaviour with another apparently irrelevant one. We agree with Dawkins (1986) that the term 'displacement' is unfortunate because the behaviour might be a relevant one misinterpreted by the observer. The behaviours are however very striking and tend to occur when the animal appears uncertain what to do or is thwarted from doing something. For example, a European Robin encountering a large beetle may dither beside it as if unsure whether to attack or flee, and then briefly preen itself. Bird feeders in which the food is temporarily unavailable induce similar behaviour. Social interactions provide many opportunities for a conflict within an individual about whether to approach or flee or for an individual to become thwarted.

Lorenz (1970) argued that 'displacement' preening during courtship had become ritualized into some of the mating displays of ducks. Such preening is often made more conspicuous by being targeted at colourful feathers. These often form a speculum—a panel on the upper surface of the secondary flight feathers—with a species-specific colour pattern (females usually have a duller version). For example, that of the Mallard is iridescent blue, bordered by white bands. Courting males expose their speculum and place their beak behind the wing as if preening. In some species the speculum is exaggerated in size because adjacent feather tracts are the same colour. For example, the iridescent blue-green greater coverts and tertials of Wood Ducks roughly double the size of the colour patch. Males of the closely related Mandarin Duck have a normally sized speculum of a similar colour to that of Wood Ducks, but have a greatly elongated central tertial feather which can be erected during courtship. The male mock-preens by placing its bill behind the sail, which then appears contiguous with the large orange patch on their cheeks, which are adorned with elongated feathers.

Turning now, to the kinds of change that occur, ritualized signals differ from cues in four main ways (Wiley 1983; Johnstone 1997): they tend to be more conspicuous, redundant, and stereotyped, and they are often preceded by alerting components. We will illustrate each of these features in turn and then assess what, if any, light they shed on signal evolution.

1. *Conspicuousness.* Signallers depend on receivers being able to detect their signals, so it is unsurprising that ritualized signals tend to be more conspicuous than cues.

2. *Redundancy.* This can involve simple repeating a signal or giving complex displays containing many elements. Both types of redundancy are well illustrated by the courtship display of Musk Ducks (Fig. 5.2). Males raise their spread tails and bend their heads back, exposing a grotesquely large flap of skin under their bill, while

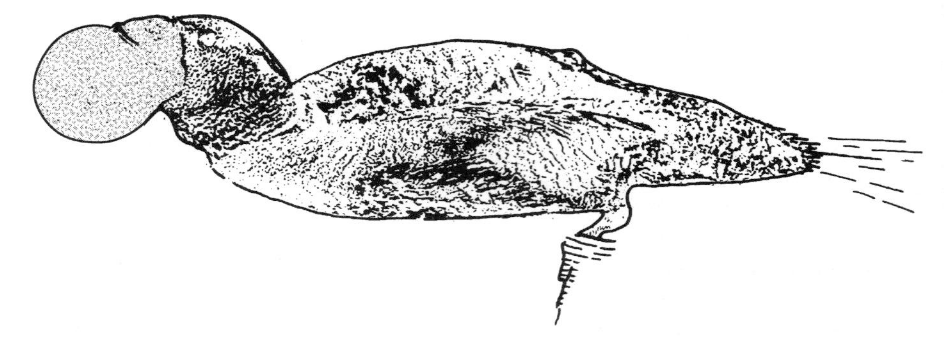

Fig. 5.2 Courtship display of male Musk Duck *Biziura lobata*; after del Hoyo *et al.* (1992).

splashing noisily around in the water with their feet. They give shrill whistles and loud grunts, the latter being accompanied by vigorous up-and-down pumping of the head with a raised crest. There may well be an olfactory component to the display as well: as noted by Darwin (1871), during the breeding season male Musk Ducks stink, hence the common name.

3. *Stereotypy.* Many displays are remarkably stereotyped. Courting male Common Goldeneyes repeatedly give Head-Throw displays. Each display starts with the head held low, almost on the bird's shoulders, and consists of three movements: (a) the neck is then extended vertically; (b) the head is flung backwards until the nape touches the rump; (c) the head moves forward to its starting position. Despite this complexity, Head-Throws vary little in duration with 95% lasting between 1.13 and 1.44 s (Dane *et al.* 1959).

4. *Alerting components.* Many visual signals start off with conspicuous movements; for example, head-bobbing in many lizards begins with fast movements of large amplitude, but is concluded by more subtle, species-specific movements (Fleishman 1992). Similarly, vocalizations are often preceded by loud calls or other sounds: in Section 7.6.3, we describe how male orang-utans sometimes introduce their calling displays by pushing a large piece of dead wood out of a tree, making a loud crash.

These four features of ritualized signals tell us rather little about signal evolution because they can be explained in several ways. Thus, Zahavi (1977) overstated the case when arguing that signals evolve as they do because ritualization increases costs and so ensures honesty. In this context, stereotypy might provide a standardized situation in which subtle differences between signallers become more obvious; an analogy would be the way in which athletes are asked to undertake the same task rather than being allowed to choose how to demonstrate their speed or strength. Ritualization could, however, have entirely different functions, to which we now turn: it may increase signalling efficacy (Section 5.2), or the ability of signallers to manipulate receivers (Section 5.3).

5.2 Efficacy

The term 'efficacy' was introduced by Guilford and Dawkins (1991), in a paper drawing a distinction between two design requirements influencing the form of signals: 'strategic design', that is, the features of a signal required to ensure honesty, and those features influencing 'the probability that the signal, once given, will reach its target destination and elicit a response at all'. They coined the term efficacy to refer to the latter features: the idea is best explained by examples.

A simple example is the observation (Morton 1975) that birds living in a forest environment tend to have lower frequency songs, a tendency that can be seen even within a species (Hunter and Krebs 1979, for Great Tits), essentially because high frequency sounds are dispersed by twigs and leaves. A second simple example is that alerting components at least sometimes really do draw attention to the message that follows, as their name suggests. For example, this was confirmed by experimental manipulation of playback for the introductory tonal elements of the subsequently trilling song of Rufous-sided Towhees (Richards 1981). Thus, ritualization does sometimes function to increase efficacy: for examples involving other aspects of ritualization, see Wiley (1983).

A remarkable example of a signal that has evolved to be effective in difficult circumstances has been described by Aubin *et al.* (2000) in the Emperor Penguin. These birds nest in the Antarctic winter. No nest is made. The single egg, and later the chick, is carried on the feet of one or other parent, while the other goes to sea to collect food. When the chick is older it remains on the ice while both parents feed. This system requires that the adult birds can find each other, and their own chick, in a colony of several thousands, when the birds are clumped together to keep warm, and in the total absence of landmarks. They do so acoustically. In birds, sound is produced in the syrinx, a two-part organ located at the junction of the bronchi. The two parts can be controlled separately. The result is a 'two-voice system'. In the Emperor Penguin, the beat generated by interaction between the two parts results in a sequence of 'syllables' of varying length, resembling a bar code, in the apt analogy suggested by the authors. Playback experiments demonstrated that, provided the two-voice system was being used, birds were able to identify their chick, or partner, at a distance. The two-voice system also occurs in the King Penguin, which has a similar method of raising its chicks, but not in nesting penguins. In this fascinating case a highly complex signal has evolved in a cooperative system to meet the requirements of efficacy.

It may require careful analysis to determine what type of signal is most likely to reach its target. Endler (1983, 1991) has emphasized that it is not sufficient that a colour pattern should look conspicuous against an artificial background to a human observer. Parrots, tanagers, orioles, and titmice can be hard to see against their usual bright background, as can a brightly coloured butterfly flying in flickering sunlight. Endler has worked on poeciliid fishes, in which males are faced with the contradictory requirements of being conspicuous to females but cryptic to predators. He treats the colour pattern of the background as consisting of a mosaic of patches varying in size, shape, colour, and brightness. A fish is cryptic if its own pattern resembles one drawn at random from the background, allowing for lighting conditions and the visual sensitivity of potential predators (e.g. two poeciliid species use red patches for communication, because their main invertebrate predators are red-blind). In guppies large and highly reflective spots are more conspicuous: they occur mainly on males from regions with low predation.

In some cases, the optimal design of a signal depends on its function. Marler (1955) described a particularly convincing example. Alarm calls can be divided into different categories according to the response to the signal. 'Flee' alarm calls, given to a cluster of individuals in immediate danger of attack, for example from a hawk, and causing all individuals rapidly to disperse, we would expect to be hard to locate, and not necessarily audible from a distance. 'Assembly' alarm calls cause hearers to assemble from a wide area, perhaps to mob a potential predator: we would expect them to be easy to locate, and audible at a distance. Figure 5.3 shows that these expectations are confirmed.

Guilford and Dawkins (1991) emphasized the importance, for some types of signal, of 'receiver psychology'—in particular, the efficacy of a signal will be increased if its meaning can be easily learnt and remembered (Roper 1990). For example, domestic chicks learn to avoid a visual signal more rapidly if it is paired with an auditory stimulus (Rowe 2002), consistent with the idea that redundancy increases efficacy.

5.3 Arms races, manipulation and sensory bias

5.3.1 Introduction

A more cynical view of animal signals was taken by Dawkins and Krebs (1978), and developed further in Krebs and Dawkins (1984). This view sees a signal, not as a means of transmitting information, but as a means whereby one animal exploits another's muscle power. The signaller is seen as 'manipulating' the receiver, and the receiver as 'mind-reading'—that is, as deducing, by experience or instinct, how the signaller will behave. The result is an evolutionary arms race. Krebs and Dawkins (1984) point out that the features of ritualized signals are those used by the advertising industry to persuade us to do something to their benefit and—all too often—to our cost.

In ESS models, it is usually assumed either that the signaller can choose one of a small number (often only two) alternative signals, or that the signal varies continuously along a single dimension. For such models, there is usually an ESS, characterized by the fact that each party to the interaction is choosing the optimal behaviour—signal or response—given what the other party is doing. But there is an alternative kind of situation in which, whatever the present state of the population, there is always a better signalling policy, and always a better policy for the receivers;

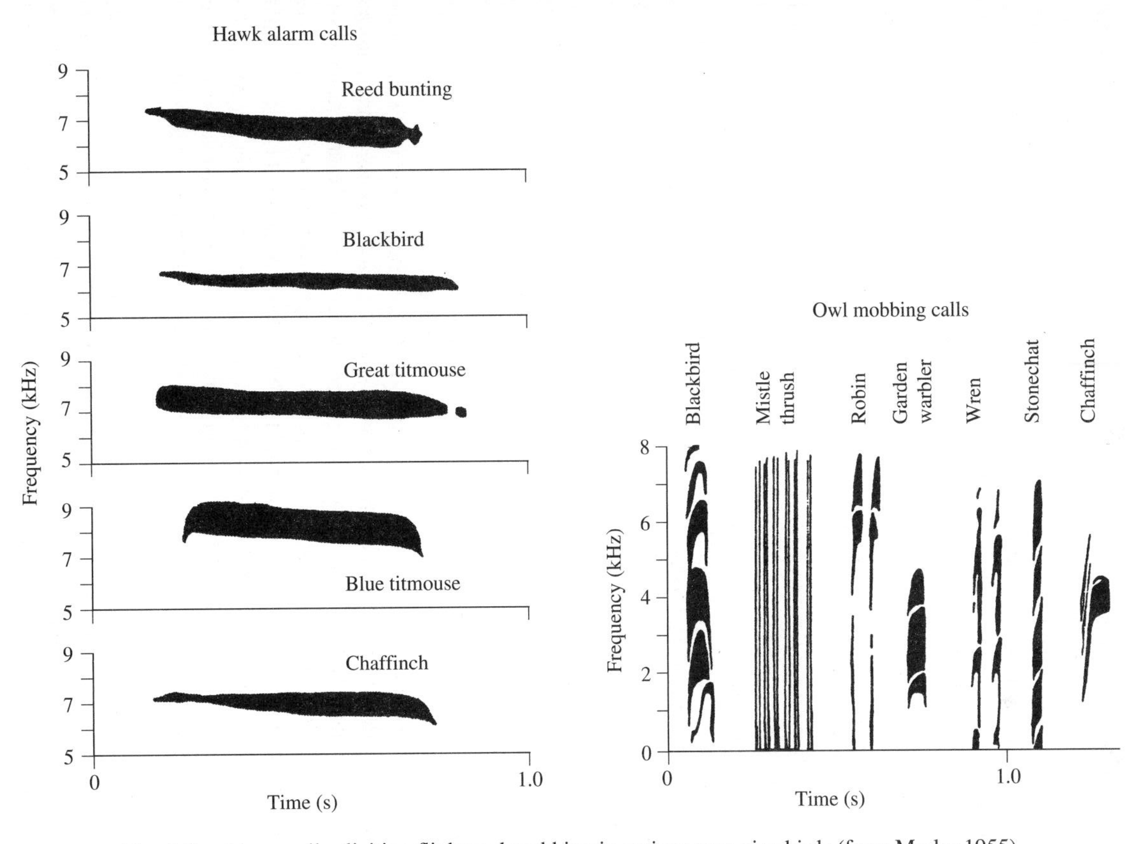

Fig. 5.3 Alarm calls eliciting flight and mobbing in various passerine birds (from Marler 1955).

there can be no stable equilibrium, and the population will evolve continuously. The crucial feature of such situations is that, whatever the current state of the population, there is always a possible signal that will extract more from the receiver than the signal that is currently being given.

We now discuss this alternative. We will do so in three stages, First, we describe a simple model that leads to continuing evolution, and an example in which it has been shown experimentally that a population is not in equilibrium. We then ask what evidence there is suggesting that, whatever signal is now being given, there is always a possible improvement, extracting more from the receiver. Finally, we discuss comparative data that point to non-equilibrium dynamics.

5.3.2 A model, and an experiment

Enquist (2002) considers the following simple model. An 'actor' gives a single signal, saying in effect 'I am here, and want something'; a 'reactor' responds by transferring a gift of value x, which can vary from 0 to 1. The actor's fitness is maximized if he receives $x = 1$; the reactor's optimum is $x = 0.5$. Thus there is a conflict of interest. In the first version of the model, Enquist assumes that, depending on genotype, the actor can choose any one of a number of unique signals, and that, again depending on genotype, reactors have a 'look-up table' indicating the response to every possible signal. Mutation of an actor causes him to choose a different, randomly chosen signal; mutation of a reactor causes him to alter the response to a random signal (usually, of course, one that is not at present being given) to a new value, randomly chosen between 0 and 1. If the number of possible signals is reasonably large, it will always be the case that there is some mutant signal that will extract a value of x close to 1 from most reactors, and that it will be some time before an appropriate mutation in reactors, giving x close to 0.5 to the new mutation, can arise and spread. It is therefore not surprising that, in simulations, actors evolve so as to extract a reward substantially larger than the value of 0.5 that is optimal for reactors. In other words, reactors are being manipulated, or exploited.

Enquist recognizes that the model is implausible mechanistically. It is not plausible that reactors should have a genetically determined look-up table for responses to many unique signals, most of which are not being received. He therefore simulated a revised version of the model in which the responses of the reactor are determined by an evolvable neural net. Behaviourally, this differs from a look-up table in that it 'generalises'; that is, if two signals are similar, then the responses to those signals will also be similar. In simulations, if both actor and reactor populations were able to evolve, the degree of exploitation was small ($x < 0.51$), although there was continuing evolution of both. However, if only the actor population could evolve, the degree of exploitation increased. Thus, although with coevolution the outcome was close to the reactor's optimum, we are not looking at an ESS; at an ESS, even if one participant could no longer evolve, there would be no inducement on the other to change.

Turning to the real world, the following experiment (Rice 1996) illustrates such non-ESS behaviour beautifully, albeit in a non-signalling context. In *Drosophila melanogaster*, male seminal fluid reduces the competitive ability of sperm from other males, thereby increasing the male's fitness. It also reduces the female's propensity to remate, and increases her egg-laying rate. However, it is toxic to females, reducing their survival. Rice allowed males to evolve in adaptation to females of a particular strain, but prevented the females from coevolving (by continuing the female strain from females that had not been exposed to the selected males). The result was an increase in the toxicity of the seminal fluid of the selected males in the unselected females. It seems, therefore, that in nature males are continually evolving in an arms race with other males, and that females respond to the changes in males so that the seminal fluid does not become increasingly toxic; if females are prevented from coevolving, the seminal fluid does become more toxic. Perhaps this will be seen as the coevolution of chemical weapons rather than of signals, but it does show how coevolution can lead to an ever-changing balance.

5.3.3 The response to novel signals

In the models described in the last section, the evolution of the population depended on the response to novel signals. If the response to a novel signal is arbitrary, and unrelated to the present signal to which reactors have had a chance to adapt, then the outcome is dramatic and unpredictable, and most of the time the signallers are exploiting the sensory preferences of the reactor. But if, as simulated by a neural net, responses are positive to novel signals that resemble to present one to which reactors have adapted, but negative to wholly new signals, then continuing evolutionary change still occurs, but the degree of exploitation of signallers is small. How do animals in fact react to novel signals?

Experiments suggest that the usual response to novel signals, rather than being arbitrary as in Enquist's first model, are more likely to be positive to signals resembling the one to which the receiver is already adapted: see Mackintosh (1974) for a review. This phenomenon, known as 'generalisation', has in the main been established by first training an animal to respond positively to some stimulus, X, and observing that its responses to new signals falls off with their difference from X (see Fig. 5.4a). However, similar results have been obtained for instinctive responses, although the experiments are harder to design and interpret, because the relevant features of the signal often concern pattern rather than some continuous variable such as intensity or pitch (Ghirlanda 2002). Note that, if the response to the initial signal X was already optimal for the signaller, then there would be no continuing evolution. But this is not always so: reactors may prefer a signal differing somewhat from that to which they have been trained (or have evolved) to respond—a phenomenon known as 'peak shift' (see Fig. 5.4b). If so, although not arbitrary, the response to a novel signal is such as to cause continuing evolutionary change. In some cases, animals may prefer signals of substantially greater intensity, a phenomenon referred to by Tinbergen (1951) as a 'supernormal stimulus' (see Fig. 5.5).

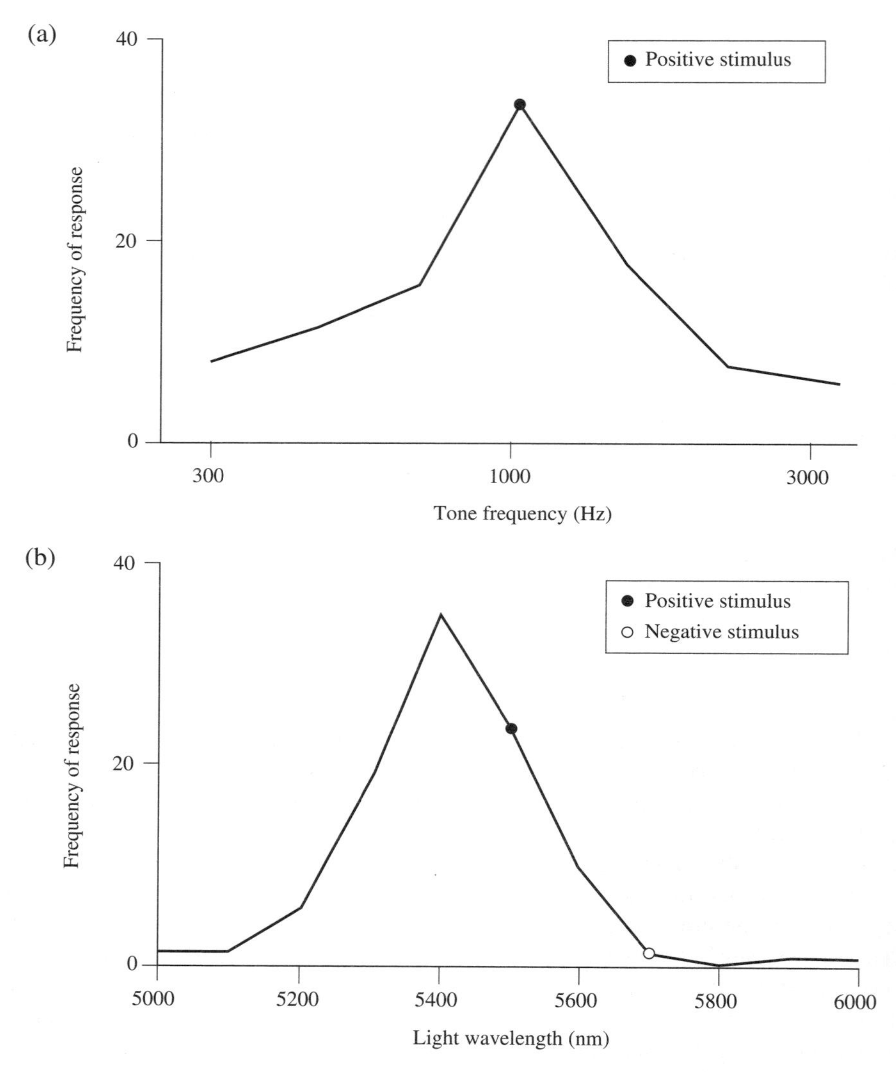

Fig. 5.4 Generalization and peak shift (after Enquist 2002). (a) Generalization: response of pigeons as a function of tone frequency. The birds had previously been rewarded for responding to a frequency of 1000 Hz (Jenkins and Harrison 1960). (b) Peak Shift: response of pigeons as a function of light wavelength. The birds had previously been rewarded for responding to a wavelength of 5500 nm, but not for responding to 5700 nm (Hanson 1959).

What of qualitatively novel signals? There is evidence for two kinds of novelty:

1. A cue that is relevant in one context evolves into a signal in a wholly different one. An example is described in Section 5.4.3: in guppies, there is evidence that orange colouration, which is used as a positive cue by both sexes when foraging, has

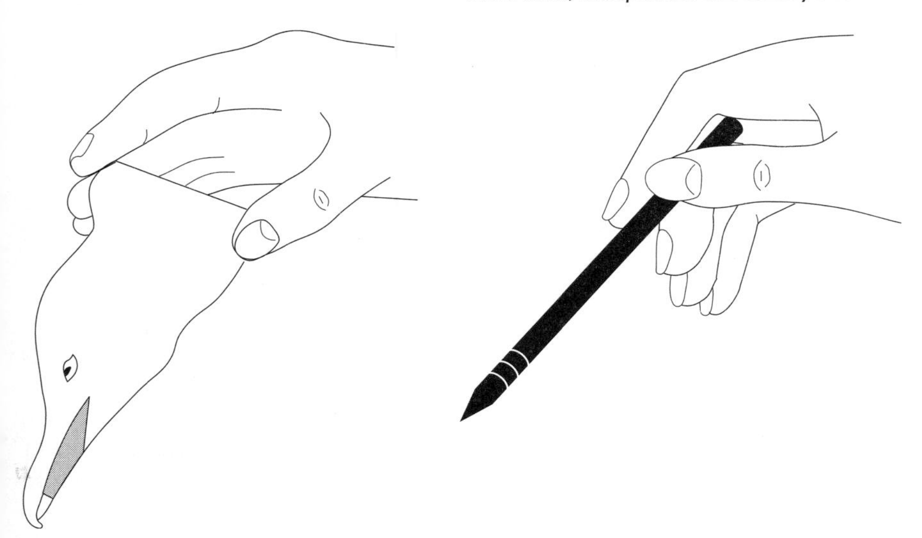

Fig. 5.5 Example of a supernormal stimulus: Herring Gull *Larus argentatus* chicks aim 26% more pecks at a thin red rod with three white bands at the tip than at a more realistic cardboard model of an adult's head; after Tinbergen (1953).

evolved into a sexually selected male trait. A similar process has occurred in water mites feeding on copepods (Proctor 1991; see Section 5.4.1).

2. A qualitative change may occur in a signal, without a change of context; for example, there may be a change in the location of a patch of colour influencing female choice in courtship. This has been most intensively studied in relation to colour rings attached to the legs of birds, perhaps because ornithologists are worried that the rings they use for identification may have unintended consequences. Zebra Finches have a red bill, brighter in the male. Males with red colour rings have been reported to be more attractive as partners, both in captivity (Burley 1981) and in the wild (Burley 1988). In a freely breeding population, red-banded males were more successful in obtaining extra-pair copulations than males with green bands, and females mated to red-banded males resisted extra-pair copulations (Burley *et al.* 1994). However, the phenomenon is controversial: Jennions (1998) failed to detect a preference for red bands in cage choice experiments. Female preference for males with leg bands similar in colour to that present elsewhere in his plumage has been reported in some other species (e.g. American Goldfinches, Johnson *et al.* 1993) but not in others (e.g. Bluethroat, Johnsen *et al.* 2000).

The phenomenon is not confined to leg bands. Zebra Finches and the related Long-tailed Grass-finch belong to a lineage devoid of crested species, yet females prefer males wearing artificial white crests (Burley and Symanski 1998). In the Zebra

Finch this might reflect transference from the white 'moustaches' of the male: the Long-tailed Grass-finch does have a white rump, but in this case transference is perhaps less plausible.

Transference can also occur when the initial preference is learnt, not genetic. Plenge *et al.* (2000) raised female Javanese Munias under two conditions. Some were raised by normal unadorned birds, while others were reared by birds with a red feather added to their foreheads. Once mature, the females were allowed to choose between a normal unadorned male and a male ornamented with three different kinds of artificial ornament, which differed from the learned red feather on the forehead. These novel traits were: (a) a blue feather on the forehead; (b) red stripes on the tail; (c) blue stripes on the tail. Females whose parents had red feathers on their foreheads preferred males with red stripes on their tail, but not males with either sort of blue ornament. Thus, there appeared to be transference of a learned preference for the colour red. This kind of transference could play an important role in the evolution of conspicuous male traits. The control females, raised by unadorned parents, showed no consistent preference for unadorned or adorned males. This shows that females could recognize males with a phenotype different from their father as conspecifics. It also suggests that selection for mating with a conspecific would not necessarily prevent the evolution of novel sexually selected traits.

Thus, it is impossible to be dogmatic about the nature of the response to novel signals. However, it seems likely that the common pattern will be of generalization, perhaps with a small degree of peak shift, resulting in continuing evolutionary change, but not of a rapidity to result in serious exploitation of the receiver of the signal. But, occasionally, non-adaptive bias in favour of a wholly novel signal may have more dramatic results, less favourable for the receiver. Note that these conclusions depend on two assumptions:

1. There is a conflict of interest between signaller and receiver over the optimal response to a signal.
2. The signal is not an index.

In the next section we discuss the empirical evidence for this type of non-equilibrium dynamics.

5.3.4 The comparative data

If the appropriate image of signalling systems is a shifting balance, rather than a stable equilibrium, there is one clear prediction. Since a shifting balance is predicted as a consequence of conflict between signaller and receiver, we would expect the form of aggressive signals to evolve more rapidly than cooperative ones. Andersson (1980), using data from Sebeok (1977), argued that this was the case. He explained it, not in terms of sensory exploitation, but by suggesting that a novel threat display is usually a reliable predictor of attack, but later is increasingly used as a bluff and so loses its effectiveness, and is then replaced by yet another new display.

We would also expect aggressive signals whose form is not causally linked to quality to evolve more rapidly than indices of RHP used in aggressive encounters. Enquist (2002) quotes aggressive signals in cichlid fishes in support of this prediction. Colour patterns are often used in aggressive encounters; these almost always differ between closely related species. In contrast, tail-beating and mouth-wrestling are indices of quality, and are shared by many species.

5.3.5 Conclusions

In contrast to the ESS picture, in which the population evolves to a stationary equilibrium, at which it would not pay either signallers or receivers to change their behaviour, given what the others are doing, the 'arms race' picture supposes that there is always a 'mutant' signal possible that elicits a more favourable response for the signaller, and concludes that, if this is so, the result will be continuing evolutionary change and exploitation of receivers by signallers. Formal models have confirmed this intuition, but have shown that much depends on how receivers respond to wholly novel signals. If responses to novel signals are such that there is always some signal possible that elicits a substantially more favourable response, then the result will, as predicted, be continuing evolution, and serious exploitation of receivers. However, experiments on the response of animals to novel signals suggests that the typical behaviour is one of 'generalization': that is, the more similar a novel signal is to the one at present preferred as a result of training (or, if responses are innate, to the signal commonest in the present population), the more favourable will be the response. If generalization is perfect, then there is no selection for change, and hence no continuing evolution. But if there is some degree of 'peak shift'—that is, if the most favoured signal differs to a small extent from the one that has been optimal during training—then there will be continuing evolution, but the degree of exploitation will be small. However, it should be emphasized that most of the evidence concerning generalization and peak shifts is based on experiments in which the responses to signals are the result of training rather than instinct. Finally, there is empirical evidence that, if receivers are prevented from evolving, serious exploitation can result.

This 'arms race' picture makes two predictions that are confirmed. First, we would expect signals used in aggressive encounters to evolve more rapidly than those used in cooperative interactions. Second, in aggressive interactions, purely conventional signals (e.g. colour patterns) should evolve more rapidly than indices of RHP, for example of size.

5.4 Sensory manipulation

5.4.1 Frogs and swordtails

A related idea has been proposed by Ryan (1990, 1998), and Basolo (1990, 1995). They suggest that a signal may evolve whose form depends on a pre-existing bias in the sensory system of receivers. The form of such signals may appear arbitrary, but it is not a necessary feature of the idea that there should be a continuing arms race between signaller and receiver.

In the Túngara Frog (Ryan 1990) the male mating call has two components, a high-pitched whine and a series of low-pitched chucks. The whine is always given and is sufficient to provoke a female response. Chucks are not always produced, but increase the attractiveness of calls to females: they are also risky, attracting frog-eating bats (Ryan 1985). Larger males make a lower frequency chuck, and females prefer lower-frequency calls. Low frequencies are perceived by the basilar papilla, and the tuning of the papilla is such that it is more sensitive to frequencies lower than that of typical male chucks; thus the female preference is probably a direct consequence of this bias in sensory sensitivity. Ryan argues that this 'sensory bias' existed in the genus *Physalaemus* before the evolution of the chuck call (Fig. 5.6). The *P. pustulosus* species group consists of two sister clades, one on each side of the Andes. Chucks are part of the mating call only in two of the eastern clade: the Túngara

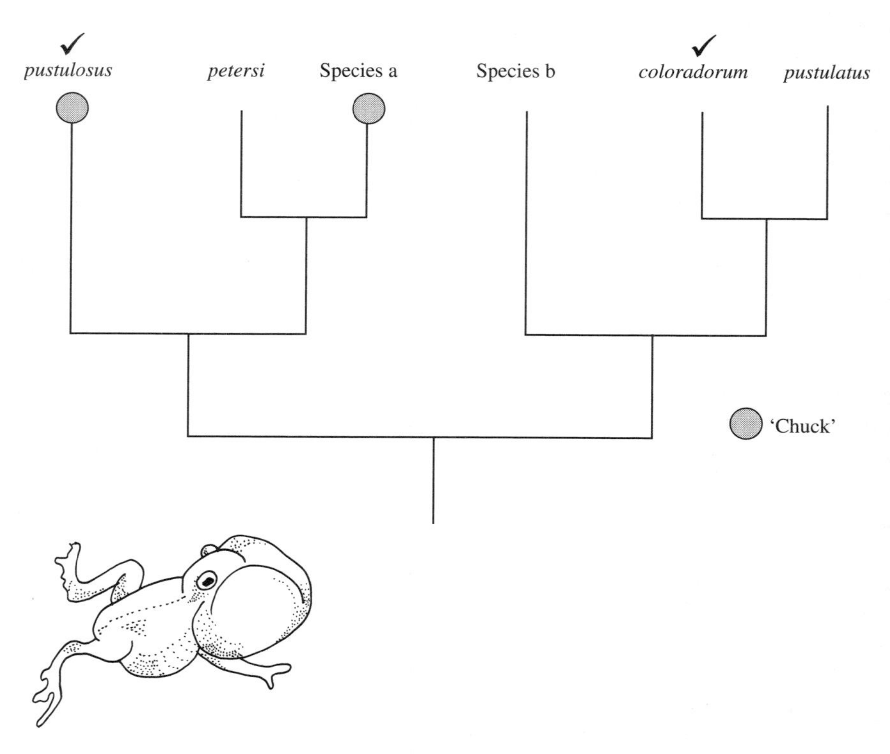

Fig. 5.6 Phylogeny of the Túngara Frog *Physalaemus pustulosus* clade, based on morphological and molecular data; after Ryan (1997). Males of the species labelled with filled circles often add a chuck to their advertisement call. Females of the two species labelled with ticks prefer recordings with chucks added to those without.

Frog *P. pustulosus* itself and an undescribed species that is a sister to *P. petersi*. Yet the basilar papilla (see earlier) is present throughout the genus, and measurement of sensitivity in a species from the western clade, *P. coloradorum*, which lacks the chuck call also revealed a sensory bias towards low frequencies. Moreover, females of that species prefer conspecific whines to which chucks have been digitally added (Ryan 1997). This suggests that the sensory bias did indeed precede the evolution of the chuck call, although an alternative possible explanation is that predation by bats has led to the secondary loss of the chuck in some species (Shaw 1995).

A similar argument has been made for the 'sword' in the fish genus *Xiphophorus* (Basolo 1990). This genus includes both swordtails, with a sword-like extension of the caudal fin, and platyfish, with no such extension. Basolo found that platyfish prefer conspecific males with a 'sword' to the same males without. If the presence of a sword is a derived character in the genus, then this suggests that the female preference for a sword preceded the male structure. Meyer *et al.* (1994) presented a molecular phylogeny suggesting that the absence of the tail in platyfish is the derived character within the genus. This makes the traits of the sister genus, *Priapella*, critical. They lend support to the sensory bias hypothesis: males lack a sword, but females prefer conspecific males with one added artificially (Basolo 1995).

Thus in *Physalaemus* and *Xiphophorus*, female preference preceded the male trait. This observation is relevant to Fisher's (1930) 'runaway' theory of sexual selection. Ryan (1998) points out that, in both the 'runaway' and the 'good genes' theories of sexual selection, female preferences evolve indirectly because they are genetically correlated with male traits that are under direct selection: the preferences themselves are not under direct selection. This is correct. For example, in the runaway theory, a female that mates with a male with, for example, a long tail does not therefore have more offspring; but the gene that causes her to do so will, in the next generation, tend to be associated with genes for long tails, and males with long tails *do* have more offspring (provided, of course, that most females prefer such males). Thus, the spread of a gene causing females to prefer males with long tails occurs because of 'genetic correlation' with long-tail genes. A problem with the theory, therefore, has always been to explain why most females come to prefer such males in the first place. Fisher himself thought that, typically, females would select as mates males of high general fitness, and that sometimes such preferences would be exaggerated by his 'runaway' process. The theory of sensory bias provides an alternative answer: the preference was originally an accident of sensory physiology. However, this does not invalidate Fisher's theory. Given an initial preference, for whatever reason, selection may lead to an exaggeration both of the preferred trait, and of the preference. Sensory bias would also help to explain why preferences tend to be for larger, brighter, or otherwise more conspicuous traits: Fisher's theory by itself predicts the evolution of tiny tails, for example, just as readily as large ones.

These two examples have in common that the signal elicits a positive response in species lacking it. In neither case is there any reason to see the signal as exploitative, or as the first step in an arms race. It is not obvious why the chuck, or the sword, should have appealed to receivers in the first place. A possible answer is that the novel trait was relevant to some cue or signal already influencing female choice. For example, female fish might favour a sword because of a pre-existing preference for large males: in other words, swords were initially exaggerators of body length. This hypothesis is not supported, however, by a study of the swordless species *Priapella omaecae*: females prefer males with a sword added, but do not discriminate between males of different size (Basolo 2002). A second type of explanation for sensory bias is that there was a pre-existing response that was adaptive in a different context. For example, water mites, *Neumania papillator*, locate their copepod prey by detecting water vibrations. Proctor (1991) showed that males attract females by moving their legs so as to mimic these vibrations.

We now discuss a context in which it seems that a signal that initially elicited an adaptive response from females can evolve into one that is effective only because females have evolved to respond to it, despite the fact that they no longer gain any direct benefit from doing so.

5.4.2 Nuptial gifts in insects

Male insects often offer nuptial gifts to females. There is a bewildering variety of systems, reviewed by Vahed (1998). The following summary is based on how females may benefit from the gift, and what the male gains from offering it.

5.4.2.1 The female benefits nutritionally

We describe two examples. Mating in the scorpionfly *Bittacus apicalis* was described by Thornhill (1976). Males capture an arthropod prey, and then attract a female pheromonally. The male presents the prey to the female, which feeds on it during copulation. After copulation is ended there is a struggle for the prey; the male usually wins, and either eats it or uses it in a second courtship. The female benefits from a large prey item through an increase in her subsequent fecundity. A second case in which females benefit nutritionally was described by Steele (1986*a*,*b*) in *Drosophila subobscura*. During courtship a male often produces on his proboscis a drop of liquid regurgitated from his crop. A starved female, but not a well-fed one, will extrude her own proboscis and take the drop. The food thus obtained increases her subsequent fecundity. Steele showed that wild-caught females usually behaved like starved laboratory females.

In both these cases there is evidence that the nuptial gift increases the female's fecundity; there is no reason to speak of 'exploitation'. But the benefits to the male are quite different in the two cases. In *Bittacus,* as in most insects in which nuptial gifts are exchanged, a female will eat and copulate at the same time. The larger the prey item, the longer the duration of copulation, and the more sperm are transferred. The male does not increase his chance of mating by his gift—the female mates anyway: he increases his fecundity. In *D. subobscura,* Steele found that starved females are more likely to mate if offered a nutritious drop: the male is increasing his mating success. It may also be relevant that female *D. subobscura* typically mate only once, so, by increasing the fecundity of the female, the male is also increasing the number of his own offspring.

5.4.2.2 Sensory exploitation

Sakaluk (2000) argues that the evolutionary origin of nuptial gifts in the cricket *Grylloides* is sensory exploitation. In these crickets, the male deposits on the female genital aperture a spermatophore consisting of two parts, an ampulla containing sperm, covered by a jelly containing free amino acids, called a spermatophylax. During copulation the female eats the spermatophylax; the quantity of sperm transferred depends on its size. There is phylogenetic evidence that this system is derived from that in *Gryllus*, in which there is no spermatophylax, and the female terminates copulation by eating the ampulla; she may then mate again with the same male. Females of three species of *Gryllus* readily eat the spermatophylax of *Grylloides* if offered it, showing that the sensory response preceded the structure. Sakaluk suggests that the system in *Gryllus* evolved into that in *Grylloides* because it pays the male to prolong copulation by offering a gift that is equally attractive but cheaper to produce.

If this interpretation is correct, then *Grylloides* has taken the first step on the evolutionary path from a nutritionally valuable gift to the sensory exploitation of the female, although it is important that the signal is by no means arbitrary in form. Empid flies (also known as Dance Flies) have travelled further along this path: see Cumming (1994) for a review. In typical Empids the males form a mating swarm, each carrying a prey item as a gift. During copulation the female eats the prey; copulation ceases when she has completed her meal. In some species the prey is lightly wrapped in silk: in others it is partly encased in a silken balloon. In *Empis geneatis,* the balloon contains a minute, dessicated prey item, useless as food (Fig. 5.7). In two species of *Hilara,* the gift consists of an inedible silk ball containing no prey at all. Although there is no direct phylogenetic evidence, it seems plausible that evolution has led from a system analogous to that of *Bittacus* to one in which the female responds to a signal, but gains no direct benefit in return. She may, however, receive benefits of the kind proposed by Fisher (1930): by mating with a male that offers an inedible balloon, she will have sons that offer a balloon, and that is what females like.

5.4.3 Further examples of sensory manipulation

Female guppies prefer to mate with males with bright orange spots. The degree of preference varies strikingly between different geographic regions, which has made guppies a favoured system for the study of mate choice, starting with Endler (1980). Until recently, the favoured hypothesis was that, by choosing males with bright spots, females choose males of high quality, for two reasons. First, the orange colouration of the spots depends on carotenoids, which must be ingested, and is therefore a measure of foraging ability. Second, the colour depends on parasite load (Houde and Torio 1992).

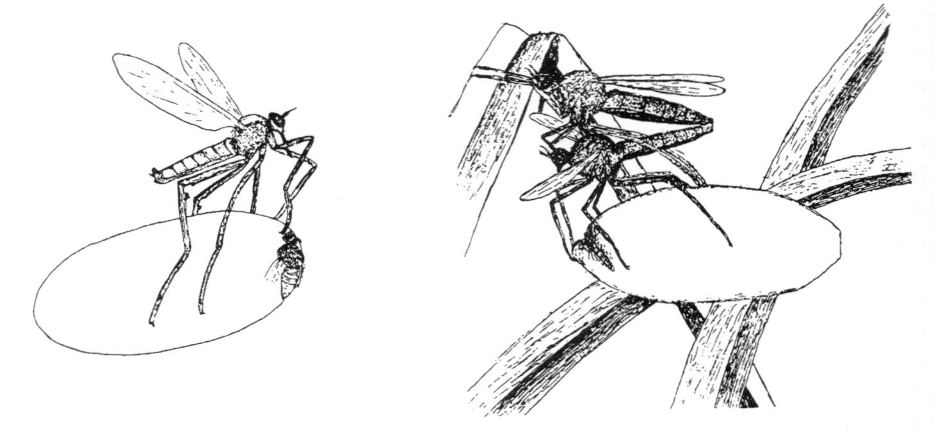

Fig. 5.7 Nuptial gift in *Empis*: the male carries a silk balloon containing a prey item (from Brown 1975).

The first reason for doubting this 'indicator' hypothesis was the observation (Grether 2000) that the strength of female preference is not greater in regions where carotenoids are hard to acquire. More recently, Rodd *et al.* (2002) have produced strong evidence for the view that males are exploiting a pre-existing sensory bias in favour of orange objects. They studied fish in the wild, and also fish derived from different regions, but raised in the laboratory on brown-coloured food. Colour preference was measured by recording the frequency with which fish pecked metal discs of various colours. In laboratory studies, males as well as females pecked at orange discs more frequently than at discs of other colours, and the degree of preference in males and females covaried in fish derived from different regions. They also found that male preference for orange discs, and female preference for males with bright orange spots, covaried geographically: astonishingly, male preference for orange discs accounted for 74% of the geographical variation in female mate preference.

5.5 Mimicry and cheating

It may be helpful to define what we mean by deception (Semple and McComb 1996). Consider a signal that is given in more than one circumstance, but always produces the same response in receivers. Receivers usually benefit from their response, but deception can occur if there is another circumstance in which the same response benefits the signaller at the receiver's expense. This definition makes no assumption about what goes on inside the brains of the signaller and receiver. Mimicry is a common form of deception, although the two terms do not cover an identical range of cases. Thus mimicry may be of a cue rather than a signal, and deception may involve an identical signaller giving an identical signal, but in inappropriate circumstances.

As an example, consider the jumping spider, *Portia fimbriata,* which
web spiders such as *Zosis geniculatus* (Tarsitano *et al.* 2000). When first
a web, *Portia* manipulates it in a way that simulates the struggles of
This usually provokes the resident to approach. But when moving close t
spider, *Portia* makes brief rocking movements, resembling the disturban
by wind, thus masking its own movements. A third signal, resembling that made by a prey insect brushing the periphery of the web, is made if *Portia* estimates that the resident spider is large.

Perhaps the classic example of mimicry is that in which a distasteful and warningly coloured prey is mimicked by an edible mimic (Batesian mimicry). Such systems are usually in equilibrium, at least in the short run, because the mimic is rare relative to the model, so that it pays predators to believe the signal. The edible African Swallowtail Butterfly illustrates this point beautifully, males are non-mimetic, perhaps because of the role of colour in mate choice. Females mimic different distasteful butterflies in different parts of Africa: in regions where models are rare or absent, the females are non-mimetic (Sheppard 1958). This is a game two can play: some predators are also mimics—for example, Painted Redstarts (Jablonski and Strausfeld 2000). These birds exploit the fact that many insects, when they detect an approaching predator, jump and fly away. The redstarts spread their tails and wings, which are decorated with white patches, and pursue and feed on the insects they flush in this way. Experiments with models demonstrated that both the movement and colouration of the wing and tail increase their effectiveness. The method works only because redstarts form only a small part of the predator guild, and flight is an effective defence against most predators.

In Batesian mimicry, it may be true both that the system is in equilibrium, and that the responses of receivers have evolved in the context of signalling. A Batesian mimic is certainly deceiving, or exploiting, the receiver of the signal. But the receiver of the signal responds in the required way to the signal because the response has evolved as an adaptive response to an honest signal by genuinely distasteful prey. Provided that mimics are not too common relative to models, the system is in equilibrium. Evolutionary change can arise from selection on receivers for the ability to distinguish mimic from model, and on mimics to make such distinction impossible: there is little reason to expect such selection to lead to ongoing evolutionary change; in particular, there is no reason to expect an arbitrary change in the signal to be beneficial. Essentially the same argument applies to other cases of mimicry. Mimicry, therefore, is a clear example of exploitation, but not of either sensory bias or of non-equilibrium dynamics.

The inappropriate use of alarm calls is another clear case of deception. Matsuoka (1980) watched Great Tits, Marsh Tits, and Willow Tits at a bird table giving hawk-alarms in the absence of predators. When the other birds flew off, the caller seized some food. Møller (1988) used playback of tape-recordings to confirm that the hawk-alarm calls of Great Tits given in the absence of a predator caused the same fleeing response as those recorded when a predator was present. Although the caller usually benefited at the receivers' expense, it probably paid birds to flee whenever they heard an alarm call: the costs of being killed far outweighed losing a few minutes of feeding time. Even so, there must be limits on how often deceptive alarm calls can be used before they lose their effect. Great Tits seem to ration their use in at least three ways. First, they did not use them when confronting an individual subordinate to themselves. Second, they were most likely to use them when food was likely to be unusually valuable (during snow storms, early and late in the day). Finally, birds gave fewer deceptive calls if food was dispersed, so that it was easier for birds to share it than if the same food had been clumped. Several other birds use alarm calls in similar ways (e.g. Munn 1986*a*,*b*; Tramer 1994).

Deceptive use of alarm calls during fights has also been suggested (e.g. Dale and Slagsvold 1995). Caution is required: the idea that the 'conflict call' of Eurasian Nuthatches was used both in fights and as a hawk-alarm (Cramp and Perrins 1993) arose because two very similar calls were confused (Matthysen 1998).

Courtship is a common context for deception. For example, female Dance Flies cannot hunt for food, and receive all their protein in the form of nuptial gifts from males. They compete for these gifts at leks which form at dusk. Experiments using model flies show that males prefer to give their gifts to and then copulate with females with extended abdomens. In some species (e.g. *Rhamphomyia sociabilis*) abdomen swelling is a reliable index of egg maturation. But before female *R. longicauda* join a lek they swallow air, inflating pouches along the edges of their abdomen. They then wrap all three pairs of legs—which are heavily scaled—around the abdomen. These two exaggerations completely mask the state of egg development, so that males are regularly deceived into handing over food to females whose offspring they are unlikely to father (Funk and Tullamy 2000). But it is not only males that are deceived in Dance Flies: as described in Section 5.4.2, these flies have long been known for a case of apparent deception in which males hand over empty silken shrouds rather than prey. It is intriguing that a system that originated as an honest trade, in which males supply food and females supply eggs ready to be fertilized, has, in different lineages, evolved into deception of males by females, and of females by males.

Adams and Caldwell (1990) studied the threat display given by the stomatopod crustacean, *Gonodactylus bredini*. Individuals in the process of moulting continue to give the display, although they are in fact vulnerable and unable to attack. The rarity of the cheating relative to the honest signal is ensured by the fact that only a small proportion of individuals are moulting at any one time. A similar case is described by Backwell *et al.* (2000) in a fiddler crab, *Uca annupiles*. Male fiddler crabs have a large claw which is used as a signal both in attracting mates and in contests with other males: in the latter context it is also used as a weapon. In most species, a male that loses a claw regenerates a new one identical in form to the original. In *U. annupiles*, in contrast, the replacement claw is lower in mass and less effective as a weapon. The authors found that males with original claws are not more likely to fight with males with regenerated claws, and females when choosing mates do not discriminate against males with regenerated claws. Surprisingly, they found that up to 44% of males in natural populations may have regenerated claws. These observations lead them to suggest that the apparent rarity of cheating may be an artefact of the difficulty, in most cases, of detecting cheats, and that dishonesty may be quite common.

This case raises a hard question. Are we looking at a stable equilibrium of the kind investigated theoretically in Chapter 2, or at a signalling system which, in this species, is on the way out? If it is really impossible for other males to distinguish between original claws and regenerated claws that are cheaper to produce but less effective as weapons, why do males produce effective but expensive claws in the first place? If the present system is a stable one, there must be some disadvantage to the replacement claw. Presumably this arises in the context of escalated fights.

6

Signals during contests

6.1 Introduction

There is an obvious question to ask about signals made during animal contests. How can such signals be reliable when the contestants prefer different outcomes? We have already discussed two possible answers to this question:

1. The signal is an index, either of fighting ability or of need for the resource being contested. It is honest because lying is impossible.
2. The signal is a handicap: in the context of animal fights, it is risky to make, and would be too costly for a low quality individual, or for an individual not in serious need of the resource.

We now consider a third possible answer to the question. This answer works only because, although the contestants prefer different outcomes, they do have a common interest—for example, in avoiding an escalated fight. Although a contestant has no stake in the survival of its opponent, it does share with that opponent an interest in avoiding escalation. Such contests have been widely studied by economists and psychologists under the title 'consensus games', but have received rather little attention until recently from biologists (but see Farrell 1996 and Rabin 1996). Such games are the main topic of this chapter.

In Section 6.2 we discuss 'badges of status'—more or less permanent patches of colour that influence the outcome of contests between individuals, although not themselves correlated with fighting ability. We discuss the circumstances in which such badges can settle contests.

Due to their permanence, badges of status cannot be used to signal temporary changes in motivation—for example, individuals may vary in hunger, and hence in their readiness to fight for a resource, but permanent badges cannot signal these differences. If contests could be settled without escalation in favour of the contestant in greater need, this would be of advantage to both contestants. But can signals lead to such a settlement? This question is discussed in Section 6.3. Rather surprisingly, models suggest that it is possible for minimal-cost signals honestly to reflect differences in need, and so lead to a consensus. However, the stability both of badges of status, and of minimal-cost signals indicating need, depends on 'punishment'. That is, it must be impossible for an individual dishonestly to signal a high degree of motivation, but then, if challenged by another aggressive individual, to retreat without cost. This assumption is crucial: punishment is discussed in Section 6.4.

For the possibilities mentioned so far, there are formal models, albeit rather crude, showing the conditions that must be met if a stable signalling system is to evolve. There is, however, one aspect of aggressive signalling that has not been successfully modelled. This is the fact that many species have a range of distinct signals, sometimes indicating varying levels of aggression, and that a contest between two individuals often consists of a protracted exchange of such signals. Obviously, such a contest cannot be modelled as a simple action–response game of the kind discussed in Chapter 2. Classical game theorists represent such a contest as a 'game in extended form', consisting of an expanding tree of possible actions and responses. Several authors (e.g. Bergstrom and Lachmann 1997; Hurd and Enquist 1998) have used such game trees to analyse interactions of the type 'A signals, B signals, A acts', but it seems hopeless to apply the method to longer sequences of acts. It would in any case be pointless, because it is absurd to suppose that an animal thinks 'I did this, and he did that, and . . . and . . . and . . . and so I will do X'. Rather, at any step, the action of an animal probably depends on the values of one or a few 'internal state variables', or 'motivations', which change during a contest according to simple rules. An explanation of behaviour then requires us to find the rules governing changes in motivation during a contest, and to show that these rules are evolutionarily stable. We are still a long way from achieving this.

We discuss the problem of extended interactions as follows. In Section 6.5.1 we discuss the empirical evidence, and discuss how far signals do fall into discrete classes, and how far they predict the outcome of interactions. We then describe two particular cases in some detail. In Section 6.5.2 we describe the fighting behaviour of cichlid fishes, and a model, the 'sequential assessment game' (Enquist and Leimar 1983, 1987) developed to explain it. The essential feature of this model is that the various actions performed are assumed to provide increasingly accurate information about fighting ability. Section 6.5.3 describes the fighting behaviour of funnel-web spiders over web sites (Riechert 1978, 1984). Behaviour, and the outcome of these fights, is influenced by ownership, relative size, and by the value of the web (known only to the owner). A model (Maynard Smith and Riechert 1984) is described that attempts to explain the behaviour in terms of changing motivation during the fight. This is a physiological explanation, not an evolutionary one, but the case is unique in that not only behavioural and ecological but also genetical information is available.

All the contests so far discussed are over indivisible resources—for example, an item of food or a position in a peck order. If the resource can be shared, the possibility for a consensus is increased. It may pay both contestants to share the resource rather than risk an escalated fight, but signals are needed to negotiate a suitable division. An obvious example is the settlement of the boundary between two territories. Section 6.5.4 describes a formal model, the 'negotiation game', of such a process.

6.2 Badges of status

Badges of status (Krebs and Dawkins 1984) are claimed to be the animal equivalent of sergeants' stripes (Roper 1986). They are patches of colour that influence the outcome of contests between individuals, even though they are not logically correlated with fighting ability (Resource Holding Potential *sensu* Parker 1974). Badges of status have always been the subject of controversy, on both empirical and theoretical grounds (e.g. Shields 1977; Rohwer 1978; Whitfield 1987) and have even been described as 'paradoxical' (Keys and Rothstein 1991). Correlations between colouration and status have been reported for many taxa including butterflies (e.g. Shreeve 1987), fish (e.g. Martin and Hengstebeck 1981) and lizards (e.g. Rand 1990). Some of the best examples of badges, however, involve conspicuous plumage features of birds, and we will describe one to illustrate some general points. Before doing so, it is worth emphasizing that plumage variation does not always correlate with status; particularly convincing cases involve Ruddy Turnstones (Whitfield 1986) and winter flocks of House Finches (Belthoff *et al.* 1994; McGraw and Hill 2000).

6.2.1 An avian example

Eurasian Siskins are highly gregarious finches. The males usually have a black bib on their chin under the bill (Fig. 6.1) but females and juveniles do not. Bibs are larger on dominants than on subordinates (Senar *et al.* 1993). Individuals prefer to feed near a conspecific rather than alone unless the other bird has a large bib, regardless of whether that bib is natural or has been painted on by researchers. A naturally large-bibbed

Fig. 6.1 Badge of status of a Siskin (from Cramp and Perrins 1994).

bird becomes more 'attractive' as a companion if it has its bib experimentally reduced in size (Senar and Camerino 1998). In these experiments, the potential companions were alone to ensure that they were not revealing their status by interacting with other birds. In addition, the birds had not previously met and so had had no chance to learn each other's status. Thus, the results demonstrate that plumage variation acts as a signal. The observation that small-bibbed Siskins defer to large-bibbed ones from a distance is striking because we cannot recall any evidence from any species that the larger-badged individual tends to win escalated fights. Perhaps small-badged birds often win all-out fights because they only escalate when highly motivated (e.g. Lemel and Wallin 1993).

Observations like those on Siskins raise a central question about such badges: why do individuals defer to others wearing a few extra coloured feathers? First, how typical are Siskin bibs? We focus on cases, such as the Siskin bib, in which badges correlate with status among individuals of the same age and sex. This excludes some of the most familiar examples of status signalling such as Harris's Sparrow (Rohwer 1977; Jackson *et al.* 1988). We have also restricted our attention to simple patches of colour and ignored more complex plumage features such as the pectoral tufts of Scarlet-tufted Malachite Sunbirds (Evans and Hatchwell 1992) and the forecrown crest of the Crested Auklet (Jones and Hunter 1999). This is certainly not because such features are uninteresting, but because simple badges present the most obvious problems about reliability.

The Siskin's bib is coloured with melanin, as are most other badges of status operating between birds of the same age and sex. Well-studied examples include the black belly stripe of Great Tits (e.g. Jarvi and Bakken 1984; Jarvi *et al.* 1987; Maynard Smith and Harper 1988; but see Wilson 1992*b*) and the black bib of male House Sparrows (Møller 1987*a*; Veiga 1993; but see Solberg and Ringsby 1997). Some badges, however, involve areas of white plumage such as the chest spots of European Starlings (Swaddle and Witter 1995) and forehead patch of Collared Flycatchers (Qvarnstrom 1997). Few badges seem to be coloured with carotenoids, although yellower Greenfinches tend to dominate duller individuals (Maynard Smith and Harper 1988). The relative rarity of carotenoid badges has been explained on the grounds that such badges are more costly to produce than the commoner melanin badges. There is however experimental evidence that the size of the black breast stripe of Great Tits is reduced by exposure to parasites in the breeding season (Fitze and Richner 2002). We do not know of any badges involving structural colours, although this may reflect biologists' neglect of such plumage features (Keyser and Hill 1999). Certainly the suggestion that structural colours reflect individual quality (Fitzpatrick 1998) requires testing: Osorio and Ham (2002) describe how to assess structural colours.

Siskin bibs seem to be rather typical badges of status in ways other than their use of melanin. First, badge size (or brightness) is positively correlated with dominance; we can think of no exceptions. Second, the badge is continually visible and is not coverable as are the epaulettes of Red-winged Blackbirds (Metz and Weatherhead 1992). However, the distinction between fixed and coverable badges is not clear-cut (Hansen and Rohwer 1986). Third, badge size is a poor indication of age-class because, although badges become larger (or brighter) at successive moults, there is a large overlap between age classes (e.g. Norris 1993; Veiga 1993). The Siskin's bib is slightly unusual because it is only present in males, a feature it shares with House Sparrow bibs (Liker and Barta 2001). Virtually all badges, however, are larger (or brighter) in males.

6.2.2 ESS models of badges

As noted above, there is surprisingly little evidence that badge size correlates with fighting ability. Maynard Smith and Harper (1988) discussed a model of badges of status in which 'aggressiveness', the willingness to escalate, was signalled. The essential assumptions were as follows:

1. Badges vary continuously from 'small' to 'big'; they are not costly to produce, and are uncorrelated with fighting ability.
2. If contestants differ in badge size, the one with the smaller badge retreats, and its opponent obtains the resource without cost.
3. If badges are similar in size, an escalated fight occurs, which is more costly for opponents with large badges.

They concluded that contests could only be settled by traits uncorrelated with fighting ability if the value of the resource was small relative to the cost of fighting. Honest signalling of aggressiveness by badges could only be stable if 'cheats', which have a large badge but do not fight, suffered costs.

This model has been criticized on the grounds that the signalling ESS can be invaded by mutants other than the cheating mutant described. Owens and Hartley (1991) considered a 'Trojan' mutant, possessing a small badge, and behaving submissively when resources are common and of low value, and aggressively when resources are rare and of high value. They concluded that such a mutant could invade, and that a non-signalling ESS would evolve. Badges of status, therefore, would have to be explained by their role in some context other than the settlement of disputes over resources. However, their model assumes that only two badges, big and small, are possible. It makes two further assumptions:

1. In a contest between two small-badge individuals, the resource can be shared, without cost.
2. If resources are rare, a Trojan mutant (with a small badge) can always win a contest against an 'honest' small-badge opponent, without cost.

These latter two assumptions are implausible: there is no way in which a contest over resources can be resolved without cost in the absence of a perceived asymmetry, arising either from distinguishable badges of status or in some other way. The problem is discussed further in Section 6.4. A continuously varying badge was an essential feature of the model proposed by Maynard Smith and Harper. A Trojan mutant does not seem to make sense in such a context.

A similar criticism can be made of Johnstone and Norris (1993), who considered a 'Modest' mutant, which has a small badge, but which behaves aggressively towards small-badge opponents. No variation in resource value is assumed. As in the model of Owens and Hartley, the authors assume that only two types of badge, big and small, are possible, and make essentially the same two assumptions: that two honest small-badge opponents can share a resource cost-free, and that a Modest mutant always wins a contest against an honest small-badge opponent, also without cost. They conclude that Modest can invade, unless there is a cost to aggressiveness in contexts other than the contest itself. They therefore suggest that badges of status may provide information that is useful in contexts other than disputes over resources, for example in mate choice. However, the model is open to the same objections as that proposed by Owens and Hartley: an essential feature of badges of status is the possibility of continuous variation, or at least of a number of distinguishable badges. The notion of a Modest mutant is hard to incorporate into such a system.

Although we disagree with the model proposed by Johnstone and Norris, their motive for proposing it needs to be addressed. Males with large badges are sometimes preferred by females either as mates (e.g. the black belly stripe of Great Tits, Norris 1990, 1993) or for extra-pair copulations (e.g. the black bib of House Sparrows) (e.g. Møller 1990; but see Whitekiller *et al.* 2000). These observations are important for two reasons. First, they show that bird colour patterns do not neatly divide into cheap melanin-based ones used in agonistic encounters and costly carotenoid ones used in sexual displays (McGraw and Hill 2000). Second, if selection can maintain a polymorphism for a large badge that is cheap to produce and uncorrelated with fighting ability, as argued by Maynard Smith and Harper (1988), why should females prefer males with a large badge? A possible explanation is that, initially, the female preference was non-adaptive, and due to 'sensory bias' (see Section 5.4). If such a bias existed, it would lead to an increase in the frequency of large-badge males, which would then be less fit than small-badge males in the context of resource competition. It might then be the case that only males of higher RHP (resource-holding potential) could afford these extra costs. Female preference for a large badge would then be adaptive, because a large badge would be a signal of male quality whose reliability was maintained by their role in fighting, as suggested by Bergland *et al.* (1996).

6.2.3 Conclusions

Badges of status are more or less permanent patches of colour that settle contests over resources of relatively low value. Both observation and theory suggest that such badges can be effective, even if they are relatively cheap to produce and are uncorrelated with fighting ability. However, theory suggests that the stability of such a signalling system against a 'bluffing' mutant, which dishonestly signals high aggression, requires that such individuals are 'punished': that is, it must be impossible to signal high aggression, but retreat at once without cost if one's opponent also signals high aggression. Empirical evidence on punishment would be of great value. Criticism of the model leading to these conclusions, and suggesting that the equilibrium could be invaded by alternative 'Trojan' or 'Modest' mutants, are flawed because they assume that there are only two levels of signal, and that two individuals both signalling low aggression can share the resource without cost. However, there is evidence in some cases that there is female preference for males with a large badge, which is, therefore, a signal that operates in two different contexts—resource competition and mate choice. We suggest that the female preference may have originated as a non-adaptive result of sensory bias, but later evolved into an adaptive preference for a reliable signal of male quality.

6.3 Can signals of need settle contests?

The problem is not an easy one. We discuss it in three stages. First, we ask how contests over an indivisible resource can be settled when there are no differences in need, no perceived asymmetries, and no conventional signals: these assumptions lead to the classic 'War of Attrition' (Maynard Smith and Price 1973). We then suppose that there are real differences in need, and hence in the value of a resource to an individual, but still no conventional signals: this is the 'War of Attrition with Random Rewards'. Finally, we ask whether, given differences in need, conventional signals can help to settle contests.

6.3.1 The war of attrition

This model assumes that pairwise contests occur over an indivisible resource of value V. There are no perceived asymmetries, for example in ownership or size, or in some permanent 'badge of status' of the kind discussed in Section 6.3. We assume that, unlike humans, animals cannot deliberately introduce an asymmetry by 'tossing a coin'. The resource value V is too small to justify an escalated fight. What can they do? A possible answer is that an animal will 'display' for a time t, and then, if its opponent is still displaying, will retreat, leaving the resource to its opponent. Clearly, it must be costly to display. We are considering a type of interaction intermediate between escalated fighting and cost-free, conventional signals, in which costs arise through the expenditure of time and energy, and perhaps the risk of injury if the display contains elements of physical fighting. If there were no cost, both contestants would continue for ever, which is absurd.

Let the cost of continuing for a time t be kt. Contestant A selects time t_A, and B selects t_B ; if $t_A > t_B$, the contest continues for a time t_B, and the pay-off to A is $V - kt_B$, and to B is $-kt_B$. What is the evolutionarily stable choice of t? Clearly it cannot be some constant time, say t_0: a population playing t_0 could always be invaded by a mutant playing $t_0 + \delta$. The strategy must be a random choice of t. It can be shown that the stable probability distribution of t is $P(t) = (k/V)e^{-kt/V}$. This display time is distributed like the survival time of a radioactive atom, as expected if the displaying individual has a constant probability of stopping per unit time. It is intuitively obvious that this must be so: a contestant that has already displayed for a time t is in exactly the same situation, as far as future gains and losses are concerned, as it was at the start of the contest. The expected pay-off to a contestant, per contest, is precisely zero: an individual would be no better off, and no worse, if it gave up immediately without a contest. This contrasts with an average pay-off of $V/2$ if the contestants could toss for it, the winner taking the resource without cost.

This model is of interest mainly in emphasizing the benefits—to the population—that would accrue if contests could be settled by some means other than attrition. We have already discussed one such means, badges of status. We now ask whether a mechanism based on differences in the value of the resource to different individuals could settle contests more profitably.

6.3.2 The war of attrition with random rewards

This model was first analysed by Bishop *et al.* (1978). It supposes that individuals vary in 'need', so that the value of the resource, V, varies. The variation is non-genetic; for example, an individual is sometimes hungry and sometimes not. As in the model just described, contests are settled by costly 'display', and won by the contestant prepared to continue for longer. No signals are exchanged. This game is difficult to analyse. It turns out that there is no 'strategy', converting a degree of need into a display time, that is an ESS. No matter what strategy members of the population adopt, there is always a mutant strategy that can invade. We suspect that the population will evolve continuously, as in the 'Rock-Scissors-Paper' game (Maynard Smith 1982; Sinervo and Lively 1996) and perhaps in the evolution of size (Maynard Smith and Brown 1986).

6.3.3 A model of conventional signals of need

What of conventional signals? Can such signals be used to settle contests, without need for prolonged and costly displays? This problem has been discussed by Enquist *et al.* (1998), developing a model first proposed by Enquist (1985). The authors first show that a conventional signal that varies continuously with need cannot settle such contests: in effect, there is no way of preventing 'lying'. They analyse the case of discrete signals. Consider the following simplified version. As in the previous section, two animals compete for a resource. Individuals vary continuously in need—some are seriously in need and others are not. Two discrete kinds of signal are available, A (value high) and B (value low). If one contestant signals A and the other B, then the former obtains the resource, and neither pays any signalling cost: in other words, the contest is settled by conventional signals. But if both give the same signal, then the contest can only be settled by a war of attrition, with victory going to the contestant willing to continue for longer. The cost of such a contest is small for two individuals signalling B, but higher if both signal A. Thus the existence of two distinct signals does enable some contests to be settled cost-free, in favour of the contestant in greater need. Contestants must still pay a cost if both are in serious need, and a lower cost if neither is in serious need. The average pay-off to a contestant is positive, and not zero as in the war of attrition.

However, as Enquist *et al.* point out, there is a difficulty. What of a 'bluffing' mutant, which signals A when in low need, but retreats at once if its opponent also signals A? Such a mutant would win against other low-need individuals without cost, and would be no worse of against an A opponent. Stability of honest signalling requires that bluffing be 'punished'. Enquist *et al.* assume that, if both contestants signal A, there is a minimum cost that both must pay, and that this cost is payed by a bluffing mutant as well as by an honest A signaller.

There is an obvious similarity between this conclusion and that reached above concerning badges of status. Conventional cost-free signals can play a role in settling contests, but stability requires that bluffing be punished. The difference between the two models is as follows. Badges of status are permanent, and therefore, cannot convey information about short-term differences in need. Badges can vary continuously, and the cost of a contest is high only if two contestants have similar badges. In effect, the role of the badge is to introduce an arbitrary asymmetry which can be used to settle the contest. The stability of such a system depends on the value of the resource being low relative to the cost of an escalated contest. In contrast, in the model analysed by Enquist *et al.* (1998), the signals can reflect immediate, short-term differences of need. Such signals cannot vary continuously, but must fall into two, or a small number, of discrete classes.

In the absence of concrete examples, it is uncertain whether conventional signals of need are in fact used to settle contests. The models are described primarily in the hope that they will stimulate empirical studies. In particular, we need to know more about the effects of variation in need on behaviour. Two conclusions of the models are important. First, continuously varying conventional signals cannot be used to settle such contests: discrete signals are needed. Second, stability against a 'bluffing' mutant, signalling high motivation but retreating if challenged, requires an 'initial cost'. In effect, this amounts to the 'punishment' of lying. The same conclusion emerges from an analysis of badges of dominance: the phenomenon of punishment is discussed in Section 6.4.

6.3.4 Conclusions

Two animals are competing for a resource. The 'null model' is to suppose that they do not differ in need (or, equivalently, in the benefit they would receive if they gained the resource), and that there is no perceived asymmetry (e.g. in ownership) that could be used to settle the contest. This leads to a 'war of attrition', involving costly contests of variable duration: at the ESS, the average pay-off to a contestant is zero. It is doubtful whether animals in fact engage in such contests. If there is variation in need, but still no signalling of need and no perceived asymmetry, this is the 'war of attrition with random rewards'. No ESS appears to exist for this game.

Can cost-free signals of need evolve? Clearly, an unfakeable index of nee stable: examples were described on p. 17. But what of a conventional sign signals can be stable, but there are two restrictions:

1. Only two, or a small number, of discrete signals are possible; two contestants giving the same signal must engage in a costly war of attrition, whose cost increases with need.
2. 'Bluffing' must be punished: that is, an individual signalling high need, but retreating at once if its opponent also signals high need, must be punished.

The need for punishment to stabilize both badges of status, and the signals just considered, is important, and requires further empirical data. In effect, what is required is that, if there is a discrepancy between an individual's signal and its subsequent behaviour, the signaller must suffer a cost.

6.4 Punishment

It is not controversial that punishment is widespread among animals. In their review, Clutton-Brock and Parker (1995) describe a number of contexts in which it occurs: the establishment and maintenance of dominance; 'theft' and 'parasitism'—for example, birds stealing food or dumping eggs on neighbours are chased: the establishment of mating bonds—males punish females that refuse mating, or wander away: parent–offspring conflict—parents punish excessively greedy young: and enforcing cooperation—for example, in both *Polistes* wasps and Naked Mole Rats, queens punish lazy workers. The list of examples they give is an impressive one. However, we are concerned here specifically with the punishment of an animal for giving a false signal: that is, for signalling that it will do one thing, and doing another.

One reason why there is little evidence for the punishment of false signals may be that it would require experimental manipulation causing a discrepancy between signal and behaviour. This has been attempted in the context of badges of status. Rohwer and Rohwer (1978) painted larger bibs on the chins of captive Harris's sparrows, and found that they were attacked by cage mates and failed to rise in status. But subordinate individuals that were painted, and also injected with testosterone to raise their aggression, did rise in status. The suggestion is that, in the first experiment, the discrepancy between signal and behaviour was being punished. However, this interpretation has been challenged (Watt 1986) on the grounds that the failure of young female subordinates to rise in dominance rank in flocks containing adult males need not be the result of punishment. The evidence for punishment is stronger for male House Sparrows (Moller 1987*b*). Fights are most frequent between individuals with badges of similar size. In experiments, a dyed bird with an enlarged black bib, and a control, were introduced to captive flocks. Dyed birds did not achieve a higher dominance rank than controls but received far more aggression from naturally large-bibbed (dominant) males.

Hauser and Marler (1993) give an interesting example of an animal being punished, not for giving a false signal, but for failing to signal when it should: in effect, it is punished for 'lying by omission'. An individual Rhesus macaque calls when it finds a source of food, thus attracting other members of the group. Different calls are given depending on whether the food is specially favoured or routine. A monkey that failed to give an appropriate call on finding food received considerable aggression from other monkeys that later discovered it feeding. This aggression can plausibly be interpreted as a form of punishment.

It is worth asking why one might expect to find punishment of false signals. In aggressive encounters, the relevant context is the punishment of an individual signalling high aggression but behaving submissively if attacked. Why should it pay a dominant to punish an opponent that is retreating anyway? One possible answer is that such behaviour is an unselected consequence of the fact that the appearance of an apparently aggressive opponent would generate in the receiver a physiological state prepared for combat. Such a state would not dissipate immediately the signaller retreated, and might lead the receiver to press home its attack. Thus, in this context, 'punishment' would be an unselected consequence of a physiological mechanism.

More generally, there are two alternative explanations for the evolution of punishment when the behaviour is costly to the punisher. The simplest is in terms of individual advantage when there are repeated interactions between the same individuals: if A punishes B today, B is more likely to cooperate tomorrow. An alternative explanation was proposed by Gintis (2000), who supposes that individuals belong to small groups, which are more likely to go extinct if the members do not cooperate. The members of a group containing 'strong reciprocators', who punish non-cooperators at some cost to themselves, will learn to cooperate, and the group is more likely to survive. Gintis shows that, if the cost of punishing is small relative to that of being punished, and if the advantages to the group resulting from cooperation are large, then strong reciprocation will evolve.

6.5 Protracted contests and varied signals

The models of signalling so far described have in the main been simple 'action–response games'—a single signal elicits a single response. Many signalling interactions, particularly in contest situations, involve a protracted sequence of signals and responses. We now discuss such interactions. In Section 6.5.1, we review empirical evidence suggesting that, in such cases, signals tend to fall into discrete classes, and often do predict the outcome of the interaction. We then describe two cases in some detail. In Section 6.5.2 we describe contests between cichlid fishes over dominance, and an interpretation (Enquist *et al.* 1990) in terms of sequential assessment of fighting ability. Section 6.5.3 describes the contests between funnel-web spiders, *Agelenopsis aperta*, over web sites, as an example of an extended contest, with multiple signals, over an indivisible resource. We show how the observed behaviour can be explained by a 'motivational' model: although hypothetical, the model is supported by genetic evidence. But, although it can be shown that the behaviour has evolved in an adaptive way in different ecological circumstances, we cannot offer a full evolutionary explanation. Such an explanation would require us to show that the motivational model is evolutionarily stable: that is, that a population adopting the strategy could not be invaded by plausible alternative behaviours.

In both cichlids and orb-web spiders, the resource being contested is indivisible, and signals are used to identify asymmetries that can be used to settle the contest without an escalated fight. In Section 6.5.4, we turn to contests over a divisible resource, taking as an example the settlement of the boundary between two territories. Unfortunately, we cannot illustrate this problem by discussing a specific example. Instead, we describe a model, the 'negotiation game'; although speculative, the model does have the virtue of making some predictions. We hope it will stimulate research.

6.5.1 Varied signals—the evidence

The nature of the signals made during protracted contests has recently been reviewed by Hurd and Enquist (1998). They confine their attention to conflicts over resources of relatively small value—for example, access to a feeder. The first question they address is the following: do the signals made form a continuously varying series, or do they fall into distinct classes? They conclude that signals are usually restricted to a set of discrete levels, an idea that goes back at least to Morris (1957). For example, the head-forward threat display seems to be ubiquitous among passerine birds (Andrew 1961). Popp (1987) and Coutlee (1967) in American Goldfinches, and Dilger (1960) in Common Redpolls, all report that the display consists of a series of discrete levels (but see Martin 1970 who describes the head-forward display in the Varied Thrush as consisting of a single graded series). Difficulties arise because many displays have characteristics that are in principle able to vary continuously, so that statistical analysis of frequencies is necessary. For example, Hurd and Enquist (1998) analyse data from Brown (1975) on crest-raising in Stellar's Jay. They show that, although crest angle can vary from 0° to 90°, actual displays tend to concentrate in one of two extreme classes (~0° and 80°).

The significance of a series of discrete classes is that it becomes possible to transmit more information. Several authors draw an analogy between bird signalling and language. The analogy should not be pushed too far—there is no reason to think that there is anything in bird signalling that resembles human grammar. But the ability of language to transmit information does depend on the fact that it consists of a series of distinct phonemes. It is interesting to ask what is the repertoire size in bird signalling. From an analysis of 26 studies involving 32 species, Hurd and Enquist (1998) conclude that a repertoire of 3–5 distinct signals is typical, and a single graded display is not.

What causes a bird to select a particular signal from its repertoire? Several authors (e.g. Popp 1987 in American Goldfinches; Wilson 1994 in Silvereyes) have demonstrated the role of external factors such as season, length of ownership, relative dominance, and sex of signaller and receiver, on choice of action. An alternative approach has been 'motivational analysis'; correlations are looked for between a signal and the next act performed by the signaller—for example, attack, stay, or flee. It is important to understand that the aim of such studies is not to distinguish between external factors and internal motivation as causes of some action. In behavioural studies, 'motivation' means the internal state(s) of an animal causing some voluntary action. If an external factor influences the choice of action, it does so (by definition) by altering the actor's motivation. Hence, if a correlation is found between a signal and the next act performed, this can show two things. First, it sheds some light on the nature of the motivation—that is, the kinds of behaviour to which the motivation may lead, for example attack or flight. Second, it shows that the motivation does not vanish immediately the signal is made, but continues for long enough to affect future actions.

The results of motivational analysis have been reviewed by Caryl (1979) and Paton and Caryl (1986). First, signals are better predictors of escape than of attack. This is not surprising. There is no point in lying if you intend to give in: a poker player may be bluffing if he raises the stakes, but there is no equivalent choice if he chucks in his hand. Second, a signal that is likely to be followed by 'attack' is unlikely to be followed by 'flee', and vice versa. This, too, is unsurprising, but it does show that the signal carries information of use to the receiver. Finally, there is evidence that the behaviour of contestants is influenced by the signals they receive. Hurd and Enquist (1998) quote 15 studies showing that signals alter the behaviour of receivers. Thus, there is communication.

Summarizing, the signals made by birds during protracted contests over resources of relatively low value tend to fall into a set of discrete classes: continuously varying signals are atypical. The particular signal chosen is a predictor of the next action performed—for example, attack, stay, or flee—and does influence the future behaviour of the receiver; communication does take place. Although, for familiar reasons, the data are most extensive for birds, it is likely that they will turn out to be true of other taxonomic groups.

6.5.2 Cichlid fishes and the sequential assessment game

The fighting behaviour of the cichlid fish, *Nannacara anomala*, has been extensively studied (Baerends and Baerends van Roon 1950; Oehlert 1958; Enquist and Jakobsson 1986; Enquist *et al.* 1990). Males engage in contests that vary in duration from a few seconds to over two hours. Two males approach one another, usually changing colour, and displaying laterally. In protracted contests, there then follow three patterns of behaviour, usually in the following sequence. First, tail-beating: the fish change position, usually orienting laterally to one another, and one fish directs a jet of water towards the other by beating its tail, the fish taking turns. Second, mouth-wrestling: the fish grasp one another by the mouth and wrestle. Either fish can release itself from a bout of wrestling at will. Finally, circling: the fish swim rapidly in small circles, attempting to bite one another. A contest may end at any stage. Each type of action may be repeated many times. Some departures from the sequence outlined do occur—for example, a bout of tail-beating may follow mouth-wrestling. A contest ends when one fish signals surrender by folding its fins and changing colour; the loser is then tolerated by the winner. The outcome of a fight determines dominance in a group.

The above description is based on the account given by Enquist *et al.* (1990). The authors observed captive fish to test the predictions of the 'sequential assessment' model (Enquist and Leimar 1983, 1987), modified to allow for the presence of a sequence of different actions; the earlier versions of the model considered only repetitions of a single type of interaction. The essential features of the model are as follows:

1. Individuals vary in fighting ability, or RHP.
2. Each interaction during a contest provides information about relative RHP, but this information is subject to error. Repetition, therefore, reduces uncertainty.
3. There is a small cost to each interaction.
4. At any time, the choice made by an individual (to give up, to continue with the same or an altered action) is determined by its internal state (or 'causal factor'—McFarland and Sibly 1975) at that time. The relevant variables that contribute to this state are an individual's estimate of relative RHP, and its uncertainty about that estimate.

Enquist and Leimar (1983) found an ESS for this game, illustrated in Fig. 6.2. In effect, an ESS is a rule for deciding when to give up, given an estimate of relative

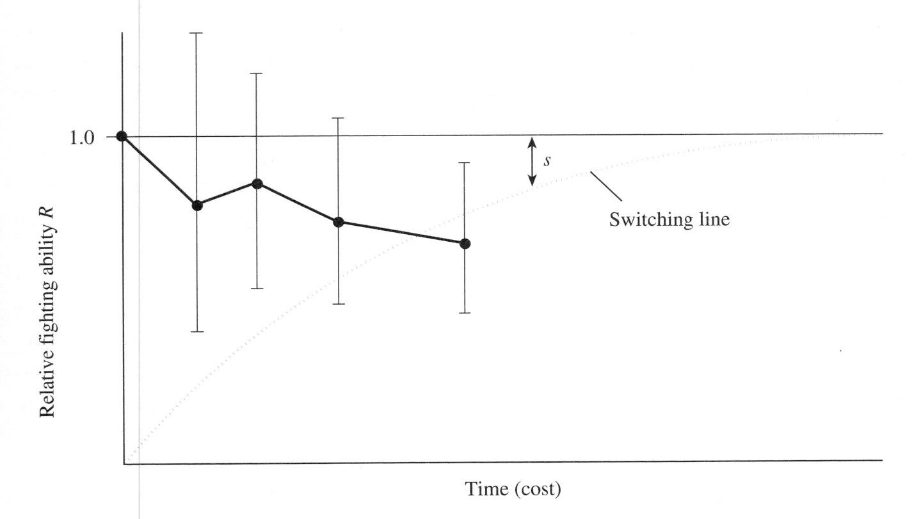

Fig. 6.2 The sequential assessment game (Enquist and Leimar 1983, 1987). The dotted line indicates a contestant's estimate of its relative fighting ability, R, after each exchange of signal, together with the uncertainty, s, of the estimate. The cost of the contest increases with time. The ESS is to follow the rule: if estimate of R falls below the 'switching line', give up. The switching line measures the degree of confidence ($1 - s$, where 1 indicates certainty) an individual can acquire about the value of R, for a given cost: it should give up if it reaches a pre-assigned degree of certainty that its fighting ability is less than its opponent's.

RHP, and its uncertainty. In Enquist *et al.* (1990) the model is extended to cover contests with several possible types of action, although the published account is only verbal. They argue that a contest will start with less costly but less informative acts, and, if such acts do not lead either contestant to give up, will progress to more informative but more costly acts.

An interesting conclusion for such models (Enquist and Leimar 1987) is that the expected pay-off to a contestant is only slightly less than the pay-off expected if the resource was divided equally between the contestants, without costs. This is in striking contrast to the 'war of attrition' game (Maynard Smith 1982), in which the expected pay-off is zero. The difference arises because, in the war of attrition, there is no assessment of RHP or of any other asymmetry, whereas in the sequential assessment game both parties gain by a relatively cheap assessment of an asymmetry in RHP. Obviously, this benefit accrues only because the value of the resource is large relative to the cost of assessment.

In testing whether the observed fights fitted the predictions of the model, the authors assume that fighting ability is well predicted by weight. Given this plausible assumption the fit is good. Although there is some unexplained variation in the pattern, duration and outcome of fights, the two main predictions were confirmed. Most contests were won by the heavier fish, and the smaller the weight difference between the fish the longer and more costly the contest.

Most contests followed the typical sequence—tail-beating, mouth-wrestling, circling. If, as seems likely, this sequence reflects an increase in the costliness of different acts, the pattern benefits both contestants. If a contest can be settled by relatively cheap actions, why engage in costly ones? But it does raise the question of how the sequence is coordinated. Hurd (1997) has shown that this coordination is achieved, at least in part, by colour displays. *N. anomala* uses two colour displays during contests, the medial line and vertical bar displays (Fig. 6.3). The ability to make these displays is not correlated with fighting ability, and is not costly. Hurd suggests that they are used in coordinating the actions of the contestants, so as to make assessment more efficient. Thus, tail-beating requires that one fish beats its tail and the other is stationary, and that the receiver is positioned relative to the sender so as to perceive the jet of water produced. Mouth-wrestling requires that the mouths of the two fish are accurately brought together. Hurd found no evidence that either colour pattern was correlated with body weight or with winning the contest. However, the median line display tended to precede tail-beating, and to be associated with the sender rather than the receiver. Similarly, vertical bars predicted the occurrence of mouth-wrestling. These associations were not perfect, but were highly significant statistically. Hurd argues that the colour displays act to decrease the cost of the contest to both contestants.

6.5.3 Spider fights, and a motivational model

Agelenopsis aperta is a funnel-web spider whose typical habitat is desert grassland. Suitable web sites are scarce, and contests take place for possession of them (Riechert 1978). The contests are usually prolonged, and consist of a series of bouts, separated

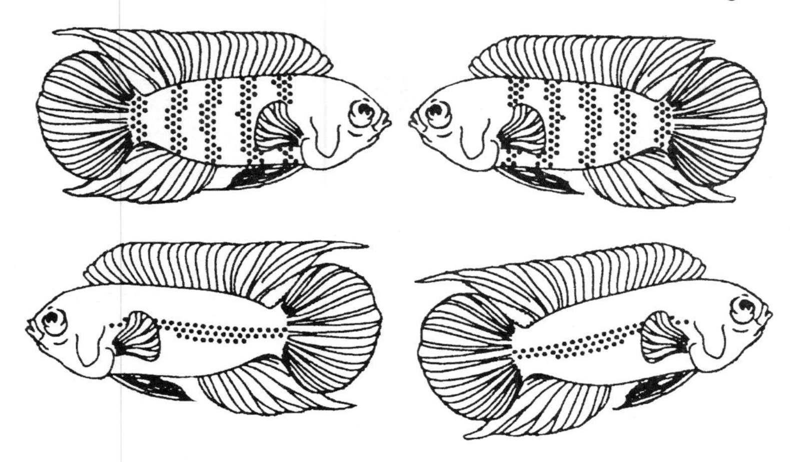

Fig. 6.3 Colour displays of *Nannacara anomala*, after Hurd (1997). The vertical bar display (above) predicts readiness to engage in mouth-wrestling; the medial line display (below) predicts tail-beating.

by pauses. Riechert recorded 33 distinct 'actions'; some of these were not 'signals' (e.g. approach, retreat), but many were (e.g. palpate legs, drum pedipalps, exaggerated lateral swinging of the abdomen). For simplicity of analysis, she grouped these actions into four categories—locate, signal, threat, and contact—with increasing levels of intensity and risk. During a bout, there is usually an escalation in intensity; a bout ends when one spider retreats. Contests end when one spider leaves the web, or when there is an escalated fight, which can result in serious injury.

The primary factor determining the outcome of fights is relative size—specifically, weight. If the difference in weight is greater than 10%, the heavier spider wins 90% of contests. An initial assessment of relative weight is made through movements on the web at a distance (locating behaviour), although it is likely that a more accurate assessment is provided later by higher intensity signals. Contests are between a resident and an intruder. If the intruder is substantially smaller, it retreats at once. Riechert (1984) showed that assessment is of weight, and that it is made through vibrations of the web and not visually, by experimentally adding a weight to the back of the smaller spider, which was often converted into the winner.

Contests between spiders differing by less than 10% in weight are usually won by the resident. The resident will contest more persistently if the web is 'valuable'; by providing food to the owner of a web, Riechert showed that value is measured by food supply. The most prolonged and costly contests, therefore, occur over valuable webs when the owner is slightly smaller than the intruder.

An explanation of these contests must therefore account for the bout structure and range of different actions reflecting different levels of intensity, cost and potential risk, and also for the influence of ownership, web value and size difference on the pattern and outcome of contests. Maynard Smith and Riechert (1984) attempted a causal explanation by adopting a theory developed by the classical ethologists (reviewed by Baerends 1975). This theory suggests that the behaviour of animals during a contest is determined at any moment by two 'motivations', or 'causal factors', which we can call 'aggression', A, and 'fear', F. If A and F are equally balanced, or almost so, the individual will continue the contest. If they are balanced at a low value, the behaviour will be of low intensity (that is, 'locate' rather than 'signal' or 'threat' in the present example); if at a high value, the behaviour will be of high intensity. If $A \gg F$, the animal will escalate to all-out fighting; if $A \ll F$, it will retreat.

Maynard Smith and Riechert modelled *Agelenopsis* contests along these lines. The model is complex. The original paper must be consulted for details, but it has the following features:

1. The next action of a contestant is determined solely by the momentary values of A and F.
2. Values of A and F alter during the contest as a result of the exchange of signals, which provide information about relative weight; hence A tends to increase in the larger spider, and F to increase in the smaller spider.
3. The effect of ownership and web value on outcome is produced by assuming that these factors influence initial values of A for the owner (but not for the intruder).

Figure 6.4 shows a simulated fight, which mimics rather well fights between desert grassland spiders. Given the complexity of the model, and the number of parameters available, the fit with observation is perhaps not surprising. However, the following points can be made in its favour:

1. The basic idea of two causal factors preceded the model, and the observations on *Agelenopsis*.
2. It is probably the simplest type of model that could explain the observations. In particular, it is hard to see how a model assuming that behaviour is determined by a single causal factor could account for them. A 'look-up table' type of model, in which the next action is specified by a knowledge of every action and circumstance preceding it, could of course be constructed, but would be highly implausible, and much more complex than the two-factor model proposed.
3. The model was constructed using only data for desert grassland spiders. However, it can also account for the genetic data now to be described, by altering only the two parameters that determine the rates of change of A and F. These genetic data are in some ways unexpected, and can most easily be explained by assuming that behaviour is determined by two genetically independent factors.

The possibility of genetic analysis arises because of the existence of a second population, inhabiting a neighbouring wooded habitat, where web sites and food are both relatively abundant. These spiders, compared to the desert spiders, occupy smaller territories, and fight less intensely over web sites. By collecting eggs in the wild and raising them in the laboratory, it can be shown that both fighting behaviour and territory size (determined by the closest distance at which a web-owner will

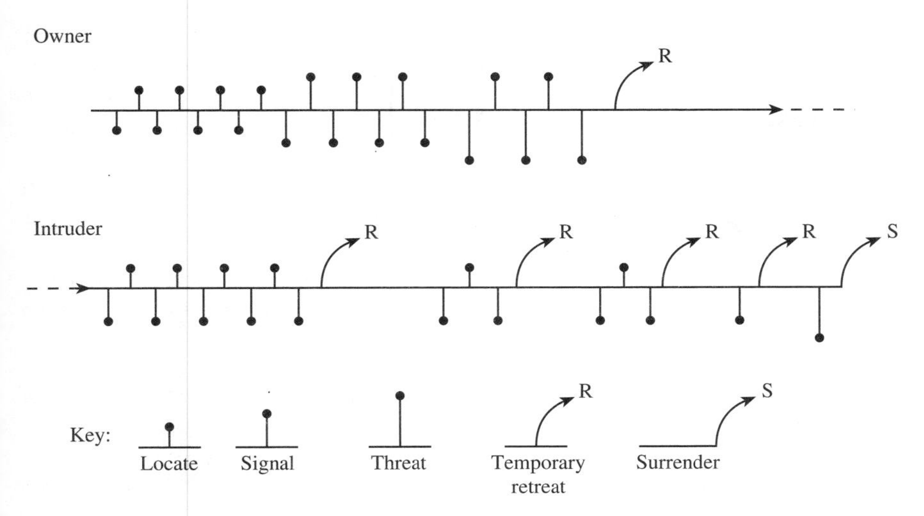

Fig. 6.4 Simulated fight between two spiders. Based on a model described by Maynard Smith and Riechert (1984). At any moment, the state of each spider is characterized by the values of two 'motivations', aggression and fear. Spiders act alternately. The model consists of two sets of rules, specifying (a) how the motivational values determine a spider's next action, and (b) how the values alter as a result of a spider's own actions and its opponent's, and of the information that those actions convey about size differences. In this particular fight, the owner is slightly smaller than the intruder. It is possible to mimic different actual fights by varying the parameters that represent external factors (ownership, size difference) and genetic factors (rules relating motivation and behaviour).

permit another spider to occupy a web) are genetic. The adaptive significance of these differences is obvious.

Surprisingly, the behaviour of the F_1 hybrids is not intermediate; they are more aggressive than either parent. Contests were longer and more costly, and often (27% of cases) ended in mortality. If behaviour is determined by two genetically independent factors, the non-intermediate F_1 can easily be explained by assuming that high A and low F are dominant. Later analysis (Riechert and Maynard Smith 1989), in which F_2 and backcross generations were studied, suggested that the locus determining aggression, A, is sex-linked.

We have discussed this case at length because it is unusual to have behavioural, ecological, and genetic data for the same traits. A 'sequential assessment' model is clearly part of the explanation—there are signals whereby relative weight can be assessed. But the situation is complicated by the fact that both ownership and knowledge of the value of a particular web also influence behaviour. It makes adaptive sense that a spider should fight harder for a web site in the desert than in the woodland, and for a web which, as owner, it knows to be valuable. But a detailed quantitative model will be hard to achieve. Perhaps more important, the existence of genetically different strains and their hybrids should make it easier to convert the notion of

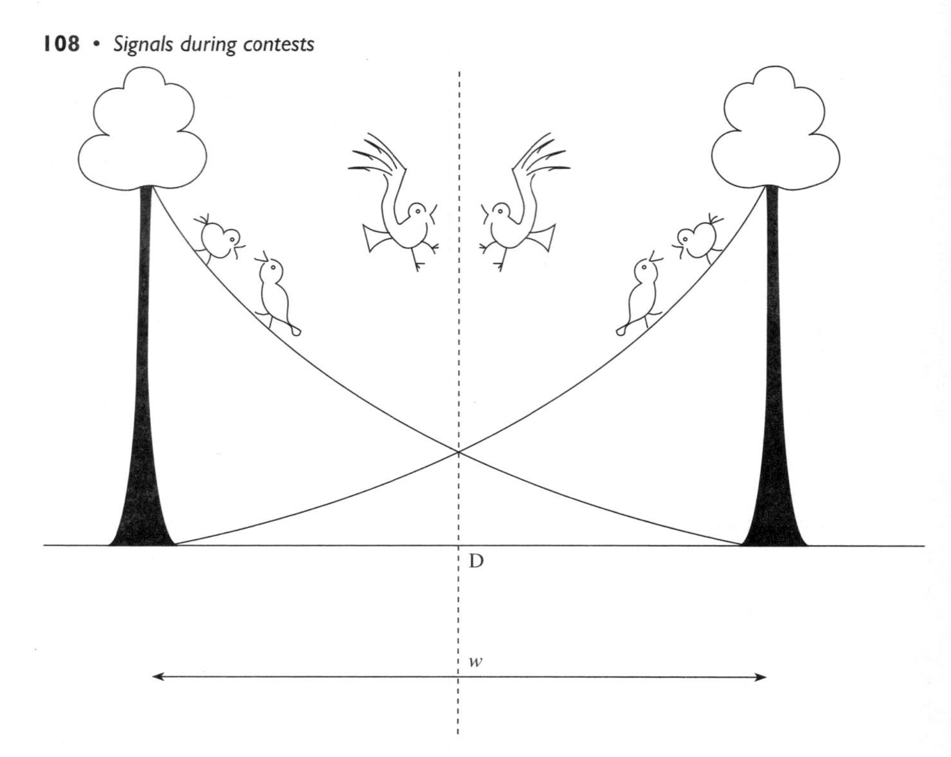

Fig. 6.5 The negotiation game; a theoretical model. There are two phases. In the first, signals are relatively cheap, and each bird signals more intensely the closer it is to its own centre. Stability requires a second phase, at the negotiated boundary, in which both birds use more escalated and costly signals.

a 'causal factor' into a defined biochemical state—most probably, the concentration of a neuro-transmitter.

6.5.4 Territorial behaviour and the negotiation game

First, as a formal model, suppose that two individuals have chosen 'centres' for their prospective territories. These might be physical features (e.g. song posts, possible nest sites) or simply areas in which they have previously been allowed to live in peace (as suggested by Stamps and Krishnan 1999). Let these territory centres be a distance w apart (Fig. 6.5). How can they settle on some point, D, marking the boundary between them? We assume that the value of a territory is an increasing function of its size.

Consider the following strategy for animal A, against animal B:

1. If you meet B at a distance x from the centre of your territory, display with intensity $f(x) = k\mathrm{e}^{-ax}$. If B displays with a higher intensity, retreat: if at a lower intensity, advance. Continue until B's display is equal to your own: accept that point as the boundary of your territory.

Clearly, if A and B adopt the same strategy, they will agree on a boundary where $x = w/2$: that is, they will divide the territory fairly. But, if signalling is cost-free, or almost so, it is equally clear that the strategy could be invaded by a mutant adopting a higher value of k. The difficulty could be avoided simply by assuming that displays are costly, but this would have the drawback that the settlement of the boundary would be very expensive, and would depend heavily on the point where A and B first met. Instead, we suppose that the contest is divided into two phases. The first, 'negotiation', follows the rule just described, with rather cheap signals. In the second phase, 'settlement', the rule is as follows:

2. If your display is approximately equal to B's, then 'escalate': that is, give a display that is costly, either because of expenditure of time and energy, or risk of injury. The cost of the escalated display is proportional to $f(x)$. If B escalates in response, the boundary is settled.

It turns out that such a strategy can be evolutionarily stable. For example, suppose that the value of a territory, V, is $x/(1+x)$, and that the cost of an escalated display is $C = ke^{-x/2}$. Then k is an evolutionary variable, measuring the cost an individual will pay to establish a territory. For a spacing between centres of $w = 2.5$, it turns out that $k = 0.57$ is evolutionarily stable: mutants that are more or less aggressive than this cannot invade. What of a mutant that displays more intensely during the first phase of negotiation, but does not match an escalated display from its opponent? Such a mutant must retreat, and will be pursued until it does match its opponent's display.

Simulation shows that a strategy of this kind can be stable, both against mutants with the same type of strategy but greater or less willingness to incur costs, and against 'liars' that display more aggressively during the first phase of negotiation, but do not match the more costly escalated display. There are features of the real world that this simple model does not include—for example, variation in distance between 'centres'. But it does have two virtues:

1. It shows the need for different levels of signal, both during the relatively cheap negotiation phase and the more expensive settlement phase.
2. It makes some predictions that are perhaps testable: that there should be a phase of negotiation in which the intensity of signals declines with distance from the centre of the territory, followed by advance or retreat as appropriate, and a phase of more costly and protracted signals at or close to the final boundary. The establishment of territories in lizards, studied by Stamps and Krishnan (1998) is a system in which these predictions could be studied further.

6.6 Conclusions

In any contest, the competing individuals prefer different outcomes, yet signals may help to settle such contests, because the participants have a common interest in avoiding an escalated fight.

The simplest situation is one in which there is some asymmetry that can be used to settle the contest—for example in ownership, in RHP or in need for the resource. Signals may not be needed if the asymmetry can be recognized without them. For example, 'ownership' can settle a contest without signals: the strategy 'if you have been in undisputed possession of a resource (e.g. a territory) for some time, fight to defend it; if not, retreat without escalation' can be an ESS if fights are sufficiently costly (Maynard Smith and Price 1973). However, there are cases in which such a strategy is reinforced by signals: the role of signals in settling contests between an owner and an intruder was discussed in Section 4.4.5.

An asymmetry in RHP usually needs to be signalled by indices of size or weapons; a number of examples of such indices have been described in this chapter and in Chapter 4. Sometimes such signals are amplified, or exaggerated. Such amplification will rapidly spread until it is employed by all members of the population; once this has happened the amplified signal again becomes a reliable index.

Unfakeable signals of need, although they do exist, are less common than indices of RHP, presumably because of the difficulty of evolving, for example, an unfakeable index of hunger. Theoretical models suggest that conventional signals of need can play a role in settling contests, provided that only a small number of discrete signals are available, and that lying signals are punished; empirical work is needed to test this idea.

If the value of the resource is small relative to the cost of escalation, contests may be settled by 'badges of status' which are effectively cost-free to produce. Such signals are not uncommon in birds, and do occur in other groups. Their stability depends on two conditions being satisfied. First, contestants with similar badges engage in an escalated contest, whose cost increases with the aggressiveness signalled by their badge. Second, individuals that signal high aggression but retreat at once if challenged by an opponent with a similar badge must suffer a cost—that is, they must be 'punished'. The punishment of false signals can therefore be an important process ensuring the reliability of signals, but both the occurrence of punishment and the selective advantage, if any, to the punisher need further study.

Many contests involve a protracted exchange of signals, which tend to fall into one of several discrete classes, rather than being continuously varying. Examples include contests between cichlid fishes over dominance, and between funnel-web spiders over web sites. In both cases, there are distinct classes of signal of differing intensity, and clear evidence that some of the signals act as indices of RHP; for example, mouth-wrestling in cichlids and vibrations of the web indicating weight in spiders. In the spider case the outcome of fights is associated with factors not associated with fighting ability; contests are more likely to be won by the owner, and owners compete more intensely for webs of high value, as evidenced by the amount of food that has been captured. It seems, therefore, that several different processes that have been analysed theoretically influence the behaviours adopted in these protracted contests, and their outcome: asymmetry of ownership, indices of fighting ability, and the readiness of individuals for whom the resource is of higher value to accept greater costs can all be involved in such a contest.

Sometimes the resource being contested can be shared; the obvious example is the division of space in territorial contests. The divisibility of a resource does not eliminate conflict—a larger share is more valuable—but it does raise problems about the role of signals in allocating shares. That signals are used in such contests is obvious, but we cannot point to empirical studies analysing their role. Instead we offer a speculative model, in the hope that it may stimulate such studies. The simplest model we can devise suggests that varied levels of signal are needed, and that the settlement of a territorial boundary may need two phases, a 'negotiation' stage involving relatively cheap signals of varying intensity, and a final 'settlement' phase involving more costly signals.

7

Signals in primates and other social animals

7.1 Introduction

This chapter discusses a number of ideas that are only rather distantly related to one another, but which have in common that they were stimulated in the first place by the study of primates, and by a comparison of human and primate signalling:

1. In Section 7.2, we discuss the role of signals in the wild, using Vervet Monkeys as our main example.

2. In Section 7.3, still using Vervets as our main example, we discuss the development of the ability to produce a signal in the appropriate circumstances, and to respond appropriately. It turns out that there is no simple answer to the question whether these abilities are learnt or innate: there is certainly an innate component, but competence is improved by experience. However, in Vervets and other primates there is little sign of teaching by parents. We compare the roles of nature and nurture in the acquisition of signalling ability in monkeys with their roles in the acquisition of species-specific song in songbirds.

3. In Section 7.4, we ask what a study of signalling can tell us about what is going on in an animal's head. By definition, a signal must, at least usually, alter the behaviour of a receiver. Many animals—not only primates—learn to make signals in particular circumstances because the result of the signal on the receiver's behaviour is favourable for the signaller. But this does not require that the signaller ascribe thoughts or beliefs to the receiver: it is sufficient that it can learn that the signal has the required effect. This is an example of the general question whether animals have a 'theory of mind'—crudely, whether they can conceive of other animals having a mind like their own. This is a much-debated issue, and we do not attempt to give a definitive answer, although hard evidence for a theory of mind in animals is hard to come by.

4. In Section 7.5, we turn to the role of 'reputation' in ensuring the reliability of signals. If animals interact repeatedly with the same individuals, and if they are able to recognize those individuals and remember their behaviour, then it is, in principle, possible that an animal that gives a dishonest signal will not be believed next time and will suffer in consequence. Evidence for such an effect is rather weak, but it should be possible to obtain it experimentally.

5. In Section 7.6, we discuss some topics not often mentioned in books on animal behaviour, In humans, behaviour is influenced not only by rational calculation of outcomes but by emotional and moral 'commitment'. Typically, such commitments are culturally acquired—they originate from rewards and punishment, from moral and religious teaching, and from ritual. However, we give one example of a human commitment that appears to be innate rather than acquired. Does commitment play a role in animal behaviour, and, if so, does 'signalling' play a role analogous to ritual and exhortation in inculcating it? We do not know. However, there are a number of displays between members of a pair, or larger group, in which all participants appear to give the same signals. It is difficult to interpret these displays as an exchange of information, particularly when they occur in larger groups. It is tempting to interpret them as the analogue of human ritual.

A final issue raised by a comparison of human and animal signals concerns the nature and origin of human language. Here we accept the view, widespread among linguists, that there are characteristics of language that are unique to humans (Section 7.7.2). In the wild animals acquire a number of distinct signs, or 'words', but they can be counted in tens, in contrast to the human vocabulary of thousands of words, made possible by the recognition of discrete sounds, or 'phonemes'. There is little convincing evidence that, in the wild, the meaning of animal utterances depends on the order in which signs are arranged, in contrast to the human ability to convey an indefinitely large number of meanings by arranging words in the appropriate sequence (syntax).

In captivity, Chimpanzees acquire a larger vocabulary, and some ability to use word order to convey or infer meaning (Savage-Rumbaugh *et al.* 1998). Such abilities are not confined to apes: indeed, they can be found in birds—African Grey Parrots can acquire an extensive human vocabulary, including words for numbers (Pepperberg 1999)—and in dolphins (Herman *et al.* 1984). Interestingly, there is evidence for some cultural variation in behaviour between groups within animal species.

However, there remains the clear distinction between human and animal language. Indeed, it is widely thought that it was the origin of language that marked the emergence of *Homo sapiens.* Some linguists have concluded that the distinction is so sharp as to require a 'macromutation' to account for it. The trouble with this, of course, is that it is not so much an argument as a cop-out. The crucial question is whether there are plausible functional intermediates between animal and human language. One linguist who has given a positive answer to this question is Jackendoff (2002); we end this book with a brief summary of his ideas.

7.2 Vervet Monkeys: a case study

The calls of Vervet Monkeys have been extensively studied, both in the wild and in captivity; the account that follows is based on Cheney and Seyfarth (1990).

Struhsaker (1967) reported that these monkeys give three different-sounding alarm calls, to Leopards, eagles, and snakes. On seeing a Leopard, and other cat species such as Caracals and Servals, a monkey gives a loud barking call; other monkeys, hearing the call, run into trees, where their agility and small size make them difficult to catch. Curiously, the call given by females, consisting of a single high-pitched call, differs from the loud bark given by males. On seeing either of the two species of eagle that prey on them, the Martial Eagle and the Crowned Eagle, Vervets give a short double-syllable cough. Vervets on the ground respond by looking up in the air, and running into bushes; Vervets in trees look up, and occasionally descend from the tree and run into a bush. Finally, a Vervet that encounters a python or poisonous snake gives a third, acoustically different alarm, referred to as a 'chutter'. Hearing a snake alarm, monkeys on the ground stand bipedally and look around; once they see a snake they may approach and mob it. Playback experiments (Seyfarth *et al.* 1980) have demonstrated that these acoustically different alarm calls elicit different and adaptively appropriate responses, in the absence of any predator, and that the nature of the response is unaffected by the length or amplitude of the alarm.

In addition to these three major alarm calls, Struhsaker (1967) reported three minor types of alarm: a minor mammalian predator alarm, given to predators such as jackals and lions that rarely prey on Vervets; an 'unfamiliar human' alarm, whose functional significance is unclear; and a baboon alarm. These three alarms are quiet, and difficult to hear or to record.

The existence of different alarm calls for different predators is not peculiar to Vervets. Zuberbuhler (2000) found that two other Cercopithecoid monkeys, Diana monkeys and Campbell's monkeys, produce acoustically distinct alarm calls to leopards and to Crowned Eagles. They form mixed-species associations, and respond to each other's calls. Birds and terrestrial predators elicit different calls in Red Colobus Monkeys (Struhsaker 1975). Ground squirrels also give different alarms to ground and aerial predators (Sherman 1977). However, their calls are not so tightly linked to the identity of the predator as in Vervets: instead, a different call is given in response to an immediate threat (usually aerial) and to a more distant one (usually a ground predator). Owings and Hennessey (1984) point out that this makes adaptive sense: ground squirrels have only one behavioural response to these calls—to escape by entering their burrow. Birds also give different alarm calls in different circumstances—in particular, to distinguish between those eliciting immediate flight, and those eliciting mobbing (see p. 75).

In addition to the dramatic alarm calls, Vervets also have a language of 'grunts' used in social communication. Grunts are given in four contexts: by a monkey when approaching a more dominant individual; when approaching a subordinate; when initiating a group movement over open plain or watching another individual initiating such a movement; and when a monkey has just noticed members of another group. Grunts given in these four contexts are hard to distinguish, either by listening or from sound spectrograms, and do not elicit any obvious behavioural responses. These observations raise the following alternatives. Is there essentially only one kind of grunt, so that any differences in response depend on the context in which the grunt is given? Or are there indeed four acoustically different signals? By recording grunts from different individuals, given in different situations, and playing back these grunts repeatedly, over several months, Cheney and Seyfarth (1982) showed that monkeys do respond differently to grunts originally recorded from individuals in different circumstances: subsequent statistical analysis revealed acoustic differences that could be responsible for the different responses. In their 1990 book, the authors comment that there are obvious advantages to giving these grunts, but that it is hard to understand why other grunts have not evolved, appropriate in other situations; for example, why not a grunt given by a mother to her offspring signifying 'follow me'? Is it possible that there is a cognitive constraint on the number of distinct acoustical signals a vervet can use?

Despite the superficial differences between the three easily recognized alarm calls and the hard-to-recognize grunts, both categories of signal convey information, including information about events external to the signaller. For completeness, we must mention a third category of call; 'chutters' which, like grunts, are hard for human observers to distinguish, but which are made in different contexts.

There is nothing unusual about the social signals given by Vervets. Juvenile Rhesus Macaques give five acoustically different 'screams' when interacting with opponents varying in rank, relatedness, etc. Playback experiments showed that mothers react differently to different classes of scream (Gouzoules *et al.* 1984). Juvenile Pigtail Macaques also give a range of acoustically different screams (Gouzoules and Gouzoules 1989), but, curiously, their 'vocabulary' is not the same as that of Rhesus Macaques; for some reason, the calls given to particular categories of opponent are different from the corresponding calls in Rhesus monkeys. We are accustomed to the rules of human language remaining the same while the meanings of particular words change: what is interesting here is that the same is apparently true of macaques.

It is interesting to compare the repertoire of Vervet Monkeys with that of Honey Bees. Seeley (1998) recognizes 17 distinct signals, chemical and mechanical, 12 given by workers and 5 by queens—a larger vocabulary than that of Vervets. Workers respond differently to a given signal depending on other information available to them. He lists over 30 cues that provide information about the state of the hive, the need for pollen, nectar and water, the needs of the brood, recognition of nest mates, and so on. Information from these cues, as well as specific signals, is required for the homeostatic functioning of the hive. It seems that bees may be integrating a larger quantity of information than monkeys, and are certainly achieving a far greater degree of colony integration. The explanation, of course, is that their ability to respond to information appropriately is instinctive, not learnt, and has evolved over many millions of years. They have more in common with the individual cells of a multicellular organism, which respond appropriately to a multitude of molecular signals, in a manner depending on their previous history.

This raises the question of how Vervets acquire their ability to communicate. Are the calls innate, or must they be learnt? It turns out that neither of these simple answers is correct: the truth is more complex.

7.3 How does the ability to signal develop?

The ability to communicate in fact comprises three separate skills:

1. The ability to produce the correct signal—for example, an acoustically correct sound.
2. The ability to signal in the correct circumstances—for example, for a Vervet Monkey, to give the eagle alarm in response to an eagle, but not a pigeon.
3. The ability to respond correctly to a signal—for example, to climb a tree in response to a leopard alarm.

Considering first the production of the correct signal, it seems that, in primates, the ability to produce an acoustically correct call depends rather little on experience. As an extreme example, Winter *et al.* (1973) found that the vocal abilities of infant Squirrel Monkeys develop normally even if they were deafened at age 4 days, or were raised for 2 weeks by a mute mother and then in isolation. Matters are a little more complex in Vervets (Seyfarth and Cheney 1980, 1986). Infants do not make alarm calls before the age of 1 month, and do so rarely when less than 6 months. However, a few alarm calls were successfully recorded from infants less than 6 months: these proved to be acoustically similar to adult calls. It seems, therefore, that practice is not needed to make perfect alarm calls. However, the observation is difficult to interpret, because infants do hear adults making alarm calls, and may learn from them. Infants give grunts at a high rate from the day of birth. These are at first acoustically abnormal, and do not come to resemble adult grunts in all respects until an age of 2–3 years.

In Vervets, then, it seems likely that the production of calls, although partly genetic, may require to be perfected by experience. The same is true of the ability to signal in the correct circumstances. For example, adult vervets give the eagle alarm only in response to raptors, and most commonly to Martial and Crowned Eagles, their main predators. Infants do not distinguish raptors from non-raptors, and may give the eagle alarm to any object in the air, even a falling leaf. Juveniles are more likely to give the call in response to raptors, but sometimes call in response to other birds. Thus there is some innate tendency to use the alarm appropriately, but discrimination becomes more precise with age. There is no evidence of explicit teaching by adults, whose treatment of infants and juveniles seems to be unaffected by whether they give correct or incorrect calls. However, it is likely that juveniles do learn by observing the behaviour of adults.

Appropriate responses to the calls of others also improves with age. In response to playbacks of alarms, Vervets of 3–4 months usually run to mother, and occasionally make potentially maladaptive responses. By the age of 6–7 months their responses are similar to those of adults. Again, there is no sign of teaching by adults. However, an infant hearing an alarm does sometimes look at an adult, and if it does so is more likely to behave appropriately, suggesting again that juveniles learn by observing adults.

Vervets also respond appropriately to several acoustically different alarm calls given by the Superb Starling. Although it is not logically impossible that this could be genetically determined, it seems far more likely that it is a learnt behaviour. Hauser (1988) observed that infants living in areas where starling calls are more frequent learn to discriminate at an earlier age.

It is interesting to compare this interaction of learning, genetic predisposition and timing with analogous processes that occur during the acquisition of song by male songbirds (i.e. oscine passerines). Males learn at least some aspects of their songs from conspecifics, and most produce highly abnormal songs if they are deprived of the opportunity to do so. A common approach to the study of song learning, pioneered by Thorpe (1958), is to rear nestlings in acoustic isolation, and to play them recorded songs. Early tape-tutoring experiments of this kind suggested that there was a brief 'sensitive period' when chicks learnt song features, and that this usually ended months before they sang themselves. For example, White-crowned Sparrows, *Zonotrichia leuciphrys*, exposed to tapes about 1–7 weeks after hatching subsequently produced the typical species song, but those hearing tapes outside this period developed abnormal songs (Marler 1970). But later work revealed that many species can learn songs outside this early sensitive period. In most of these cases (e.g. Chaffinch; Slater and Ince 1982) they learn both as recently fledged juveniles and in their first spring as they start to sing. In some species, songs can be acquired throughout an individual's life (e.g. European Starling, Eens *et al.* 1992).

When given a choice of tape recordings, naive individuals usually imitate conspecifics. This seems to reflect an innate bias rather than an inability to produce heterospecific songs. For example, White-crowned Sparrows will learn a wide variety of songs as long as those songs start with the species-typical whistle (Soha and Marler 2000). However, there must be more to be discovered about the constraints on learning in this species. Male Mountain White-crowned Sparrows have an innate bias to learn songs of their own subspecies. But these constraints are not rigid, because White-crowned Sparrows will learn the songs of completely unrelated species if they are allowed to interact socially with them but not with conspecifics (Baptista and Morton 1981).

Brood parasites, which are raised by heterospecifics, must have some way of recognizing adult conspecifics if they are to mate or defend territories. It is therefore often assumed that both the production and recognition of their vocalizations must be innate. However, it is not necessary for the whole song to be innate (Soha and Marler 2000). It is sufficient for the young bird to recognize some 'password' innately: sounds heard associated with the password can then be learnt. For example, in Brown-headed Cowbirds the password seems to be their chattering call (Hauber *et al.* 2001). Not only can male Cowbirds learn song elements from one another, but they can also modify their singing in the light of the responses of females, who prefer songs resembling those from their natal areas. In addition to influencing the retention or deletion of song elements by males, females also stimulate improvization and influence the rate and timing of song development (West and King 1988; Smith *et al.* 2000).

Finally, turning to humans, it is clear that the meanings of human words are culturally determined. However, the meanings of some human signals appear to be innate. In particular, the significance of some facial expressions are culturally universal

(Ekman 1992). A general agreement has been found between people from a range of Western and non-Western cultures when shown photographs of Caucasian faces expressing various emotions, and asked whether they indicated enjoyment, anger, fear, sadness, disgust, or suspense. This agreement was shared by natives of New Guinea, who had no previous exposure to Western images, for example in film or television. New Guinea faces expressing these emotions were correctly identified by Western observers. These findings confirm predictions made by Darwin (1872).

7.4 Questions about what is going on in an animal's head

We now turn to a very different set of questions. What do the signals given by primates, and their responses, tell us about how monkeys think? This is easier to ask than it is to answer. When we see an animal do something, it is tempting to assume that it is thinking as we would if we behaved in the same way. But, as we explain in the context of alarm calls, this need not be so. Sometimes, however, by appropriate use of playback experiments, we can get answers.

7.4.1 Do signals convey information about the external world?

Some signals convey information about the signaller. For example the black and yellow stripes on a Cinnabar Moth caterpillar carry the message 'I am distasteful', a fact about the signaller, not about the world external to the signaller. But other signals do carry such information: for example, a bird alarm call carries the message 'there is a predator close by'. So what is the argument about?

It is about what, if anything, goes on in the mind of the receiver of the signal. Consider the alarm call. It is possible that a bird is genetically programmed to fly up into a bush when it hears such an alarm. It does not have to visualize the predator, or indeed to have any thoughts at all: the response could be automatic and unthinking. Alternatively, a bird hearing the alarm call may visualize a flying predator, and respond as it would do if it had itself seen the hawk. It has acquired information about the world external to the signaller. Which is the case?

To be specific, when a Vervet Monkey hears a Leopard alarm, it climbs a tree. Does it do so because it forms an image of a Leopard in its mind and behaves accordingly, or because it follows the behavioural rule 'when you hear that call, climb a tree'? We know that a Vervet will behave appropriately when it hears a Leopard alarm, even when no Leopard is present. But what is going on in its head? Cheney and Seyfarth (1988) attempted to answer this question by habituation experiments. If an individual repeatedly hears a recording of a Leopard alarm of another individual, she soon habituates and ceases to respond. If she then hears a recording of the same individual's eagle alarm, she responds as if an eagle had been sighted. It is as if she has concluded 'there is no Leopard', rather than 'X is a liar'. This interpretation was confirmed by the following playback experiments. There are two acoustically different calls, a 'wrr' and a 'chutter', that are made in response to the presence of a rival group. If a monkey is habituated by repeated playbacks of another individual's 'wrr', she does not respond to that individual's chutter. She transfers the habituation, because the two calls have similar meanings, despite their acoustic difference; it would be hard to interpret these experiments without mental representations. Summarizing their work, Cheney and Seyfarth (1996) wrote 'Vervet Monkeys, therefore, appear to interpret their calls as sounds that represent, or denote, objects and events in the external world'.

7.4.2 Do signallers intend to alter the behaviour of receivers?

Møller (1988) reported that over half the alarm calls given by Great Tits in winter flocks were given when no predator was present (see p. 87). Subordinate birds gave false alarm notes in the presence of either dominants or subordinates, whereas dominant birds gave false alarms only in the presence of other dominants. Møller argues that subordinate birds gave false alarms to drive away more dominant individuals and thereby gain access to food. The explanation is plausible (but for an alternative explanation for the high proportion of false alarms, see Haftorn 2000). It requires only that individual birds learn that giving an alarm note increases their access to food: the calling bird does not have to think 'if I give an alarm, other birds will think that there is a predator and fly away'. An even more cautious interpretation is that the behaviour is not learnt at all: it is innate in all situations in which calling has been selectively favoured in the past.

In this example, then, there is no need to assume that an animal ascribes thoughts and beliefs to others. Humans certainly do: we would not be writing this book if we did not think that it will alter the reader's thoughts and beliefs. What of other primates? Note that we are now asking a question about what goes on in the signaller's head, not the receiver's.

It helps to start with a summary of Dennett's (1987) classification of 'intentionality':

1. *Zero-order intentionality.* The signaller holds no beliefs or desires: a black and yellow caterpillar is a likely example.
2. *First-order intentionality.* The signaller holds beliefs, but not beliefs *about* the beliefs of others. A Great Tit giving an alarm note believes that there is a predator (if the signal is honest), or that the alarm note will increase its access to food (if the signal is a lie), but in neither case need it have any beliefs about what other Great Tits are thinking.
3. *Second-order intentionality.* The signaller ascribes thoughts and beliefs to the receiver.

The existence of zero-order and first-order intentionality in animals should not be controversial. We will discuss two topics relevant to second-order intentionality: first, the influence on the signaller of the presence, nature and knowledge of potential hearers of the signal; and second, the difficulties that have been met with in the design and interpretation of experiments aimed at showing that animals are able to ascribe beliefs to others.

Vervets do not call when alone. Animals other than primates may also fail to give alarm calls when there is no selective advantage in doing so because they are alone or with unrelated individuals: for example, ground squirrels (Sherman 1977), Downy Woodpeckers (Sullivan 1985), and chickens (Marler and Evans 1996). This shows that animals may be aware of the presence of other individuals before giving an alarm, but does not require that they ascribe thoughts to others. The fact that vervets continue to call when all others in the group have seen the predator and are calling suggests that they do not.

Baboons, when travelling through wooded areas, give loud barks, audible for up to 500 m (Cheney and Seyfarth 1996). These calls may function to maintain contact between members of the group. The intentional interpretation is that baboons call in order to inform others about the location of the group: it would predict that a baboon would answer the call of another, even when itself at the centre of the group. However, 66% of barks given by females were not given in response to calls from others, but followed one of their own contact barks. Females did sometimes answer playbacks of barks given by relatives, but they did so only if their own position was peripheral. The authors conclude that 'baboon contact calls appear to reflect the signaller's own state and position rather than the state and position of others'.

To summarize, although animals are influenced, when signalling, by the presence and relatedness of potential hearers, they do not seem to be influenced in their signals by the knowledge that hearers might be supposed to possess (e.g. a monkey already giving the alarm call can be supposed to know that a leopard is present).

So, what is the evidence that signallers ascribe beliefs to others? The topic has been discussed mainly as an aspect of what has been called 'Machiavellian Intelligence' (Byrne and Whiten 1988). This concept includes a general and a more specific thesis. The general thesis is that the evolution of primate and human intelligence has been driven by the need to cope with life in a social group: this is intriguing, but beyond the scope of this book. The more specific thesis is that group-living primates have been selected to deceive other group members, and that this requires that they develop a 'theory of mind': that is, that they are able to ascribe beliefs to others, and attempt, in their signals, to alter those beliefs. There is general agreement that primates do sometimes send signals which cause others to behave as if they had been deceived. However, as we pointed out when discussing the use by birds of a hawk alarm call when no hawk is present, this does not require that the signaller ascribes beliefs to others: it is sufficient that the signaller learns by experience that the false signal has the desired effect on the receiver's behaviour.

The development of an ability to ascribe beliefs to others—often referred to as a 'theory of mind'—has been studied in human children. In an early experiment, Wimmer and Perner (1983) arranged for children of varying age to watch the following brief story. A character called Maxi puts some chocolate in a drawer and leaves the room. While he is away, his mother takes out the chocolate, uses some of it for cooking, and replaces it in a different drawer. Maxi then returns. The watching child is asked 'in which drawer will Maxi look for the chocolate?' A child aged 3 years will indicate the second drawer. A child of four knows better, and indicates the original drawer: she ascribes a belief to Maxi, and understands how he acquired it. Many experiments with a similar logic have confirmed the finding that this ability is fully developed in children aged 4 years.

What of primates? Whiten (1997) reviewed recent experiments attempting to answer the question 'do primates ascribe beliefs to others?' Such experiments have proved difficult to design and to interpret. One difficulty is that apes lack an ability present in children: they cannot talk, and so it is harder to discover what is going on in their heads. A fair summary would be that the experimental evidence suggests that higher primates are inferior to a 4-year-old child in their ability to ascribe beliefs, as revealed by the 'Maxi test' just described.

7.4.3 Conclusions

There is no doubt that some animal signals do potentially carry information about the external world, and that receivers of such signals respond in a way that would be appropriate if they had acquired that information. It is much harder to decide in particular cases whether the receiver in fact acquires the information, or whether it merely responds appropriately. In a few cases, however, there is evidence that the receiver's knowledge of the world is altered when it receives the signal. Playback experiments on Vervet Monkey alarm calls suggest very strongly that such knowledge has been acquired.

Similarly, there is no doubt that signals alter the behaviour of receivers: they would not be signals if they did not. But in no case is there clear evidence that a non-human signaller has 'second-order intentionality': that is, that the signaller ascribes thoughts and beliefs to the receiver, and intends that its signal should alter those beliefs.

7.5 Social reputation and the honesty of signals

7.5.1 Introduction

In social interactions, the honesty of signals can be ensured by 'reputation': there may be an immediate advantage from giving a dishonest signal, but this is more than counterbalanced in future interactions if the signaller acquires a reputation for dishonesty (Slater 1983). It is useful to distinguish four possible patterns of behaviour, and the cognitive abilities required in each case.

7.5.1.1 Immediate punishment of dishonesty

We have already discussed such behaviour in the context of 'signals of intent' (Section 6.5). Honest cost-free signals of intent can be evolutionarily stable provided that lying is punished immediately. Such behaviour requires no special cognitive skills: there is no need for individual recognition, or memory of past behav

All that is required is an instinctive recognition of a mismatch between signal and behaviour, and an immediate response to this mismatch. It would be inappropriate to use the term 'reputation' for such behaviour.

7.5.1.2 Direct reputation

Individual A lies to B today: B does not believe A's signal tomorrow. Such behaviour does require recognition of individuals. It also requires that A's past behaviour be remembered by B, or at least that B stores a general attitude, positive or negative, towards A. It is entirely plausible that primates should have this degree of cognitive ability. It is known that individuals remember the behaviour of other members of the group, and modify their own behaviour accordingly (Tomasello and Call 1997). For example, a male seeking allies will take into account the status of a potential ally relative both to itself and to a potential opponent. Direct reputation is the main topic of this section.

7.5.1.3 Indirect reputation

C observes A lie to B today: C does not believe A tomorrow. This requires not only individual recognition and memory of past behaviour, but also that an animal can correctly interpret a signal directed at a third party. In a review of the topic of reputation, Silk (2001) writes that we do not know whether primates are influenced in their interactions by observation of such third party interactions, but suggests that it should be possible to design experiments that would address the question. Such experiments would be of great interest.

7.5.1.4 Reported reputation

A lies to B today; B tells C about A's lie; C does not believe A tomorrow. Such behaviour is common in humans, but it is doubtful whether the communicative skills of other primates are sufficient.

7.5.2 A model

Can cost-free models of 'intent' be evolutionarily stable? If signaller and receiver place the possible outcomes of the interaction in the same order of preference, the answer is certainly yes. Even if they differ in their order of preference, such signals can still be stable if the two individuals have a common interest, for example in avoiding an escalated fight (coordination games), provided that there exists some asymmetry, known to both contestants, that can be used to settle the contest (see p. 37). We now consider a third possibility, that stability can arise, in the absence of an overriding common interest, if the same pair interact repeatedly.

This process has been modelled formally by Silk *et al.* (2000). They consider a simple action–response game. In a single meeting, the actor may be 'peaceful' or 'hostile'. If the actor is peaceful, both participants benefit by engaging in a social interaction. A hostile actor would also benefit from an interaction, but the responder would do better to flee, or otherwise avoid interaction. Can a minimal-cost signal of the actor's state be evolutionarily stable? If only a single meeting occurs, no such ESS is possible: if the responder believes the signal, it would pay a hostile actor to signal 'peaceful', and the system would collapse.

If there are repeated meetings, however, and if the responder avoids interactions with an individual who in the past has signalled 'peaceful' dishonestly, then honest minimal-cost signals can be stable. In other words, 'reputation', or memory of how an individual has behaved in the past, can stabilize honest minimal-cost signalling, even if, in a single meeting, the participants rank the different outcomes in a different order, and in the absence of any common interest. This conclusion is intuitively obvious, but a model is important because it clarifies precisely what assumptions are necessary.

7.5.3 Evidence for direct reputation

In primates, a number of quiet and apparently cost-free signals of intent have been reported. They fall into two classes: 'grunts', short, harsh sounds through open lips, and 'girneys', soft, low-frequency chewing noises accompanied by rapid lip movements. Both function to signal friendly intent. Thus, they differ dramatically from the signals of intent given by elephants in musth, which signal aggression, and whose high cost ensures their accuracy (Section 3.2.2). No comparable signals of high aggression have been reported from primates.

The account that follows is based on Silk (2001), and is concerned in the main with Rhesus Macaques and Chacma Baboons. In both species, one or other of these signals of friendliness is given when a female approaches a second, nursing, female, with the intention of handling the baby. In fact, this class of interaction was the stimulus for the model described in the last section. As in our own species, females without a baby often display a strong desire to approach and handle other's infants. The reason for this desire is not obvious. Although infants are occasionally handled roughly, it seems implausible that the behaviour evolved because the female aims to injure another female's infant in order to increase her own future infant's chances of survival in competition with existing infants. Nor does the idea that it provides inexperienced mothers with practice in handling infants seem to be in accord with the facts (e.g. adult females are as likely to handle other's infants as are sub-adult females). Silk (1999) is driven to conclude that the habit confers no direct selective advantage, and is an unselected by-product of the fact that females that are highly responsive to infants make good mothers—a rare example of a spandrel.

What, then, is the evidence that grunts and girneys are reliable signals of friendliness, and that they are believed because lies are punished? As we will see, there is clear evidence on the first point, but little on the second. Rhesus Macaques, studied in Cayo Santiago, gave both grunts and girneys. A dominant female, approaching a subordinate with an infant, was less likely to behave aggressively, and more likely to groom, if she called during the approach: she was also more likely to handle the infant, but less likely to handle it aggressively. Females were more likely to display aggression when approached if the approaching female did not call (Silk *et al.* 2000).

Chacma Baboons, studied in the Moremi reserve in Botswana (Cheney and Seyfarth 1997; Rendall *et al.* 1999), use grunts as signals of friendliness. Surprisingly, former opponents are more likely to interact non-aggressively immediately after an aggressive encounter than at other times. These reconciliations are facilitated by grunts, indicating that the signaller's intentions are now benign. The effectiveness of grunts in bringing about reconciliation was demonstrated by playing a tape of the aggressor's grunt to her former victim. As in macaques, a female that called when approaching a mother with an infant was less likely to handle the infant roughly, and more likely to be allowed to handle the baby.

Thus, the evidence that these soft calls are reliable signals of friendliness is strong. However, there is little evidence that the reliability of these signals is ensured by reputation, in part because lying signals have rarely been observed in monkeys. Silk (2001) comments that it should be possible to design experiments to measure the extent to which a monkey's behaviour towards a particular partner is influenced by previous interactions with that partner (direct reputation), and by information concerning her interactions with others (indirect reputation). Such experiments would be well worth doing.

7.6 Emotional commitment

7.6.1 *Cultural and innate behaviour*

People often do something, not because they calculate that it will pay them, but from some inner conviction that they ought to do it. As a result, game theory has proved to be a better predictor of animal than of human behaviour: the ESS concept works because animals do behave so as to maximize their fitness, whereas the similar, and earlier, concept of a Nash Equilibrium (developed to explain human conflict behaviour on the assumption that humans are rational) often fails because humans are influenced by irrational convictions. Many such convictions are 'cultural': they are acquired by social conditioning and by moral and religious teaching, or, if one happens not to share the conviction, one may prefer the term 'indoctrination'. Some, however, may be innate: we describe below the phenomenon of 'altruistic punishment', in which an individual punishes a social transgressor at some cost to himself, a pattern of behaviour which is observed in people of different cultural backgrounds, and which may well be partly innate.

Noone has ever suggested that animals are 'rational': the appearance of rationality arises because, if they behave so as to maximize their fitness given what others are doing, their behaviour will often seem rational. For example, a common behaviour in contests over an indivisible resource is the 'bourgeois strategy' (Maynard Smith and Price 1973): 'if owner, fight hard: if not owner, display only'. This behaviour does not require that an animal has an innate sense of the rights of property, still less that it calculates that, if everyone adopts the bourgeois strategy, all will be better off. It is sufficient that it follows the simple rule 'fight hard for a resource only when you have held it for some time'. In signalling, an important type of behaviour that may well be innate is to punish liars; that is, 'if there is a discrepancy between X's signal and his behaviour, punish him' (see pp. 99). This is similar to the 'altruistic punishment' observed in humans: both are probably innate, may incur a short-term individual cost, yet, as explained below, both could evolve by selection between small groups; and, of course, both can be important in ensuring the honesty of signals.

On several occasions in this book, we have used the word 'innate' for patterns of behaviour. This word is easily misunderstood. Here, by saying that a trait (including a pattern of behaviour) is innate, we mean only that its development is unaffected by variations in the environment, or learning, occurring normally in the population: of course, all traits can be affected by sufficiently drastic environmental change. The important point is that, if a trait is innate in this sense, then the strong expectation is that there has been selection favouring its appearance, despite environmental change, just as we expect that a uniform morphological trait has been favoured by selection.

There is a curious asymmetry in our attitude to humans and other animals. In humans, we have no difficulty in accepting that particular patterns of behaviour, including some deeply irrational ones, are the result of cultural conditioning, but are reluctant to believe that behaviour can be innate. In contrast, we readily accept—once we have learnt from the ethologists—that complex animal behaviour can be innate, but are reluctant to accept that cultural conditioning plays a role. This reluctance is perhaps justified in some cases (but not always—for example, see pp. 117 for the acquisition of bird song, and pp. 131 for cultural differences between chimpanzee groups). Yet, there is one possible parallel between humans and other animals which has hardly been discussed, but which we think might repay study. Human actions are often influenced by powerful emotions, which have been aroused by 'ritual' (music, dancing, eloquent speech, dressing up, etc.). There are some peculiar animal 'signalling rituals', which we discuss below under the heading 'mutual displays', that appear to have the effect, not of exchanging information, but of inducing a common emotional state in the participants.

7.6.2 'Altruistic punishment' in humans

Experimental economists and psychologists have spent much time watching their students playing games for money. They discovered that much behaviour was apparently 'irrational': players are willing to forgo financial rewards to gain less tangible ones. In particular, it appears that humans are predisposed to cooperate with others, and to punish non-cooperators, even if punishing is costly and cannot be justified by any expectation of reciprocation in the future. This conclusion emerged from a number of experiments in which small groups, usually students, play games for money (see Fehr and Gachter 2002 for a recent experiment and Ledyard 1995 for a review of earlier work). Players do not know each others' identities, and know that they will not meet again. Each player can choose to 'cooperate' or 'defect'. It always pays an individual to defect, but, if more players cooperate, everyone is better off. So, from a rational point of view, an individual is best off if he defects, and everyone else cooperates. After the choice between cooperate and defect, each player is told what each other one did, and is given an opportunity to 'punish' him: this action is costly for the punisher, but imposes a greater cost on the punished. Since the players will not meet again, there is nothing to gain by punishing a defector in the hope that he will behave better next time.

It turns out that most players report feeling angry with defectors, and accept an opportunity to punish them although, if their only aim was to maximize their profits, they would not. However, once players learn, in a series of games with different partners, that punishment is a likely consequence of defection, they are more likely to cooperate in future games. The phenomenon has become known as 'altruistic punishment'. It is altruistic because it costs something, with no expectation of future reciprocation, but, if punishing is common, cooperation spreads and everyone benefits. It seems that punishing is caused by a feeling of anger. Essentially the same phenomena have been observed in subjects from a wide range of cultural, educational, and ethnic backgrounds. It is hard to resist the conclusion that its basis is in part innate. It makes sense of a number of features of human behaviour—for example, the overwhelming need felt by the relatives of a victim of a crime to see the perpetrator punished.

A plausible explanation for the evolution of such behaviour is in terms of selection for group survival. A model proposed by Gintis (2000) supposes that individuals belong to small groups, which are more likely to go extinct if the members do not cooperate. A group containing 'strong reciprocators', who punish non-cooperators at some cost to themselves, will learn to cooperate, and is more likely to survive as a group. Gintis shows that, if the cost of punishing is small relative to that of being punished, and if the advantages to the group resulting from cooperation are large, then strong reciprocation will evolve.

An alternative, and perhaps simpler, explanation is that humans have an innate tendency to behave in a way that favours individual survival in a society in which most interactions occur with the same partners: if so, cooperation pays the individual. The behaviour carries over into the highly artificial experimental situations in which an individual does not interact twice with the same partner, so that punishment is 'irrational'.

7.6.3 Mutual displays

There are displays in which two or more individuals display simultaneously, using the same actions:

Mutual displays between members of a pair

Huxley (1930), in one of the pioneering papers in ethology, described the displays performed by a mated pair of Great Crested Grebes on lakes and rivers. These include a 'weed dance', in which the pair submerge simultaneously, collect a bundle of weed in their bills, rise to the surface, swim towards one another until face to face, and shake their heads from side to side while still holding the weed. Western Grebes and Horned Grebes have an even more spectacular finale, in which the two partners appear to run across the surface of the water with arched necks, and then simultaneously dive below the surface: these displays sometimes involves two males rather than a mated pair (del Hoyo *et al.* 1992). As Huxley pointed out, mutual displays, such as weed dances, are found mainly in species in which both sexes invest heavily in the young: they occur particularly in species such as most geese, swans and storks that pair for life.

Communal displays in groups: the Haka phenomenon

Before a match, the members of the New Zealand rugby team line up on the half-way line and perform a Maori dance known as the Haka: they jump up and down, shout, and make aggressive gestures. Such group ceremonies are observed in other animals—for example, in several species of canid prior to hunting. The signals made combine aggressive and begging elements. Those of the Silver-backed Jackal have been described as a 'harmless form of aggression' (van Lawick and van Lawick-Goodall 1970). In contrast, the ceremonies of African Hunting Dogs (described by the same authors as conveying the message 'I submerge my identity') resemble begging and those of Golden Jackals involve much mutual grooming. An association of communal hunting with group displays is not confined to mammals: for example, it occurs in both Pelicans (del Hoyo *et al.* 1992) and Double-crested Cormorants (Glanville 1992).

Some group displays precede dispersal or migration. An example is that of 'high-flying' by Bearded Tits. In late summer, flocks begin calling softly while still in their reed-bed habitat. As the calling becomes more frenzied, they fly up to height of 100 m or more while still calling loudly, only to plunge back to the reeds. Just occasionally, part of the flock peels off and flies out of sight. This is presumably the start of one of the occasional disruptive movements which are so typical of this species (Cramp and Perrins, 1993). A more familiar although less dramatic example is the gathering of Swallows on telegraph wires prior to migration, accompanied by short flights and much twittering.

In some cases, the occasion for such a group display is still obscure. For example, in the marching display of flamingos, a densely packed group of birds marches very fast, abruptly reversing direction at intervals (Fig. 7.1). The plumage is ruffled, increasing its apparent pinkness, and the birds jerk their heads from side to side (Oglivie and Oglivie 1986). This behaviour is most developed in the Lesser Flamingo, in which marching groups can contain hundreds or even thousands of birds (del Hoyo *et al.* 1992). A large group size seems to be needed for marching to occur: small flocks of captive Lesser Flamingos will only march if a large mirror is added to their pen (Pickering and Duverge 1992). Marching is often interpreted as a means of synchronizing breeding, and pairs of birds do sometimes separate from the group and display with each other. There is, however, no temporal association between marching and

Fig. 7.1 Marching and head-flagging display of the Lesser Flamingo, after Ogilvie and Ogilvie (1986).

nesting (del Hoyo *et al.* 1992), in contrast with other group displays which are most frequent in the Greater Flamingo during the last 2 months up to egg-laying (Cramp and Simmons 1977). Similarly, suggested associations between marching and other behaviours such as feeding or dispersal are not supported by data. It therefore remains unclear what function marching serves.

Another enigmatic group display (Reynolds 1965) is observed in chimpanzees. One West African word for such displays is *kanjo*, or 'carnival'. A group of adult males generate a tremendous noise by screaming, 'pant-hooting', and beating on the buttresses of certain trees, notably ironwood trees, with their hands and feet. In the Lunyoro language of Uganda, the word for Chimpanzee is *kitera*: 'it beats'. They climb and swing about in trees, and shake and break branches. Such displays may last for more than an hour. They have been observed in response to rain, and after a successful hunt. They also occur when two communities meet: in such cases, it is possible that they reflect between-group aggression, a genuine Haka.

7.6.4 The interpretation of group displays

Since, in these displays, the participants simultaneously perform the same acts, it is hard to see what information is being exchanged, or why so much energy is expended. We suggested (p. 52) that mutual displays between members of a pair may provide information about 'athletic ability', and hence act as indices of fitness in courtship: but such an explanation cannot be given for post-pairing displays, or displays in larger groups. It seems as if individuals are inducing in themselves a psychological state conducive to further cooperation—they are 'psyching themselves up'. It would be interesting to know, if one of a pair fails to participate in the display, or, having participated, fails subsequently to cooperate, whether it suffers any 'punishment'.

The nearest we have to an example is a description by Lorenz (1970) of the behaviour of a captive pair of storks, composed of a female White Stork, *Ciconia ciconia,* and a male Black Stork, *C. nigra.* Both species have a conspicuous 'Up-down' display which is used in a variety of contexts. The main sound component of this display in White Storks is a loud mechanical bill-clattering, but among Black Storks is a hissing whistle. Lorenz reports that the captive pair 'repeatedly exhibited mistrust and fear' while displaying together, and that 'the female often seemed to be on the point of attacking a male partner when he was not prepared to rattle'.

However, it cannot be the case that mutual displays in large groups are maintained by punishment of non-participants. Their most obvious characteristic is the apparent excitement of the participants, and the most obvious effect is to change the psychological state of the participants. But at present it is best to admit that these are signals whose function we do not understand.

We now return to the displays made by Chimpanzees in response to rain, because we want to discuss the implications of the fact that such displays are sometimes performed by a single male, not by a group. The earliest description of a 'rain dance' is by Goodall (1971). As rain began to fall, the group of Chimpanzees she was watching, of seven adult males and a few females and juveniles, left the tree in which they were feeding and climbed a ridge. The storm then broke, with thunder and torrential rain. One of the adult males stood, swaggered from foot to foot, gave the 'pant-hoot' call, charged at full speed down the slope, and climbed into a low branch of a tree. Successively, he was copied by the other six males, who added to his actions by tearing off a tree branch and carrying it down the slope. When all had completed this routine, they plodded back up the slope, and charged down it again. This continued for 20 minutes. The females and young watched the performance but did not participate.

Such 'dances' are infrequent, but not confined to the Gombe chimps. Takahata (1990) reports 15 such charging displays in the Chimpanzees of the Mahale Mountains, usually in a stream bed and/or in heavy rain. Displays were performed by adult males, the most dominant starting the display in 9 out of 15 cases. The displays have been filmed in the Gombe by Bill Wallauer. In an unpublished account, he reports that the displays are in response to heavy rain, or to encountering a waterfall. There may be several male participants, or only one. He describes one particular display, performed by Freud, the alpha male. Freud started his display with the typical rhythmical swaying, and by swinging across the fall on vines. At one point he stood at the top of the waterfall, and rolled rocks one at a time down the face of the fall. Finally, he swayed slowly on vines down the falls and settled on a rock about 30 feet downstream (adopting, as shown on the video, a posture curiously reminiscent of Rodin's Thinker).

What is going on? The rain dance differs from the pre-hunting displays described above in that it does not prepare the participants for a joint enterprise. It seems, rather, that it may have something to do with the establishment or reinforcement of dominance. The males seem to be exaggerating their own powers by associating them with the forces of nature (what a pity that Freud was not named after the Old Testament prophet, Elijah, who acquired a reputation for an ability to call down fire and rain from heaven). The readiness of male Chimpanzees to recruit outside resources to reinforce their displays of dominance is illustrated by the history of Mike, the alpha male of the Kasakela community in the Gombe from 1964 to 1970 (Goodall 1986). Mike's tenure of this position for such a protracted period is puzzling. He was a relatively small male, and reluctant to be involved in a fight. He was never observed to attack another male during his 4-month struggle to achieve the alpha status. His success depended on his ingenuity in display. In particular, he habitually picked up two empty petrol cans, and charged towards other males banging them loudly. When deprived of access to cans, he sought other human artefacts, or natural objects such as palm fronds, to emphasize his displays.

Orangutans also recruit resources to exaggerate their displays. Males push large snags from trees to the ground before giving long calls (Galdikas 1979). The tremendous crash produced by up to a tonne of dead wood hitting the ground makes a great alerting component (see Section 5.1) and may also be an index of strength. Males sometimes call after a nearby branch falls naturally, although it is unclear whether they are associating themselves with the noise or whether they treat the noise as a possible intruder (Galdikas 1995).

Recruitment of natural resources to exaggerate displays is not peculiar to apes. For example, Lesser Spotted Woodpeckers drum on telegraph poles next to large cracks to improve the resonance (Cramp 1985), in a signal resembling the drumming of Chimpanzees. Is there, then, anything special about the recruitment of external resources by apes? Perhaps there is. When a Woodpecker drums on a tree, the signal it produces is an exaggeration of a sound it makes when feeding: it is easy to see how the signalling behaviour could have originated. This is less obvious in the case of chimps drumming on trees. However, drumming is at least universal among chimps. In contrast, the actions of Freud were perhaps idiosyncratic, and of Mike certainly so. Ape signals that exploit nature can be individually acquired. In this as in other respects, apes are intermediate between humans and other animals.

We do not want to draw any firm conclusions from this review of mutual displays, and of individual displays that recruit external resources to exaggerate the signal. Our aim has been to draw attention to a fascinating group of facts which do not have any simple functional explanation, and which would reward further investigation.

7.7 Human language

Language is the crucial difference between humans and other animals. In place of some 30 or so distinct signals in wild animals, which can increase up to several hundred with human teaching in captivity, humans have a vocabulary of many thousands of words which, using syntax, can be arranged into sentences capable of conveying an indefinitely large number of meanings. This provides a second system of 'heredity' linking the generations, able to support rapid and continuing cultural change. With language, genetic change ceases to be the main basis of change: history begins.

In this final section, we first briefly review the extent of cultural change in animals, specifically in chimpanzees. We then describe the peculiarities of human language, and list a number of reasons for accepting the opinion, first proposed by Chomsky (1965), and now accepted by most linguists, that the ability to learn to talk is not an aspect of general learning ability, but a genetic competence peculiar to language. Finally, we ask how animal signalling could have evolved into human language.

7.7.1 Cultural inheritance in Chimpanzees

How far can behavioural differences between groups be socially transmitted without the medium of language? Boesch (1996) makes a strong case for the existence of such differences in Chimpanzees. There is no doubt that there are group differences. For example, the custom of cracking open nuts with a stone hammer is confined to chimpanzees in the Western-most forests of West Africa, although nuts and suitable stones are available in other regions where the chimps do not crack nuts. The difficulty is in interpretation: is the custom culturally transmitted, perhaps by imitation, or is there some unidentified ecological factor that makes individual learning easier in the West? The 'individual learning' explanation is unlikely: naive chimps in Zurich zoo were provided with nuts and hammers, but none learnt to crack nuts. There are other such differences. Both Gombe and Tai chimps feed on driver ants by poking a stick into the nest and feeding on the ants that crawl onto it. Not all chimp populations feed on ants in this way. More relevant is the fact that the methods used differ. Tai chimps employ a short stick and only one hand: Gombe chimps use a longer stick, and both hands. The Gombe method is some four times as efficient, as measured by ants eaten per minute. It would be hard to explain the persistence of an inefficient method among Tai chimps if learning was individual.

More relevant for the present book are differences in methods of communication. Since the meanings of the signals appear to be socially conditioned, one is driven to conclude that transmission is partly social. Boesch gives two examples. 'Leaf-clipping' was first described in the chimps of the Mahale Mountains. Holding a stiff leaf, a chimp will tear off the leaf blade with its incisors, making a conspicuous ripping sound. A similar behaviour has been observed in other populations, but it is not universal—it has been recorded only twice in the Gombe. Mahale chimps use it mainly as a courtship display: in Bossou it is seen mainly in infants at play. There are also cultural differences in the signals used by young males to attract the attention of oestrus females. The Mahale chimps use leaf-clipping. In contrast, young Tai chimps attract females by knocking with their knuckles on the trunks of small saplings. Conventional differences of meaning of this kind are likely to be transmitted culturally. The range of variation is trivial compared to that seen in humans, but it does exist.

One final example is so remarkable that we cannot resist reporting it, although it has not led to a stable culturally transmitted signal (Boesch, 1991). Tai chimps travel in groups through the forest, following a fixed direction. Occasionally an adult male will drum loudly on a tree: this drumming may serve to keep the group together. However, Boesch observed that when the alpha male, Brutus, drummed on a tree, the group sometimes changed direction. Study for over a year finally revealed that Brutus' drumming carried three distinct messages:

1. By drumming consecutively on two trees, within 1 or 2 minutes, Brutus indicated a change of direction, specified by the direction between the first and second tree.
2. By drumming twice on the same tree, Brutus indicated a rest period, of about one hour.
3. The two messages could be combined: Brutus first drummed twice on one tree, and then, within a few minutes, drummed once on a second tree. This indicated, 'rest immediately: then set off in the indicated direction'.

Brutus finally ceased using this code, perhaps because poaching had reduced the size of the group. No comparable signals have been observed elsewhere. It is as if we were observing the potential birth of a new signalling system, sadly aborted.

7.7.2 The peculiarities of human language

Most humans have a vocabulary of some tens of thousands of words. These differ from animal calls in being composed of a fairly small number (often in the range 30–40) 'phonemes'—sounds roughly corresponding to the letters of the alphabet. These phonemes are learnt, not innate, and differ between languages. Individuals learn to classify the sounds they hear into one of the locally standard phonemes. It is intriguing that in the linguistic system, as in the genetic one, a large vocabulary (words, proteins) depends on a digital 'alphabet'.

More important, the rules of syntax enable us to arrange a finite number of words in linear order to form an effectively infinite number of sentences. It is not the case that all human languages have the same grammatical rules—the same ends can be achieved by different means, although there are presumably some constraints on what constitutes a learnable grammar. Chomsky argues that the acquisition of syntax requires that humans have a special competence, which he calls 'universal grammar', enabling them to learn how to convert meaning into a string of words, and vice versa. There are a number of reasons for thinking that this competence is peculiar to language:

1. We acquire rules that we cannot consciously formulate. For example, consider the two following sentences:

How do you know what he saw?

What do you know how he saw?

Any native English speaker knows at once that the first is grammatical, and the second is not. But only a professional linguist could formulate a rule that makes this so.

2. We learn grammar with rather little reinforcement. A child would be scolded for saying 'I don't want any fucking soup', which is grammatical but impolite, but

probably allowed to get away with 'I don't want no soup'. In other words, we appear to learn complex rules with too little feedback.
3. There is a critical period for learning language. A child deprived of linguistic input until age 10 or so never learns to talk normally. Genie, a girl shut up alone by her father at 18 months, and rescued at age 13, acquired a vocabulary but never a full grammar: her speech was limited to sentences like 'want milk' or 'apple-sauce buy store'. We will return to such 'protolanguage' later.
4. Physical injury to the brain of an adult can affect linguistic competence and general intelligence separately. Damage to Broca's area can destroy grammatical competence, with rather little effect on intelligence. In contrast, damage to Wernicke's area results in 'fluent aphasia': the patients utter sentences that are syntactically almost correct, but which convey little meaning to the hearer.
5. There is some evidence that a single gene mutation can have specific effects on grammatical competence. Gopnik (1990) described a dominant gene segregating in an extended family: individuals with the gene were unable to change the form of a word to express meaning (e.g. singular/plural, present/past). The work has been criticized on the grounds that some affected individuals showed other symptoms, including poor hearing and articulation. But this seems an unreasonable criticism: a gene change which affects grammatical competence is likely to do so by affecting brain function, and such a gene is likely to have multiple effects. Bishop (2002) reviews recent progress in the field. The most exciting discovery is that the mutation discovered by Gopnik is in the gene FOXP2, a gene also present in the mouse, that is expressed in the developing brain, and codes for a transcription factor (Lai *et al.* 2001). This is what one would expect: genes that regulate development do so by producing proteins that switch other genes on or off. As yet, nothing more is known of the effects of FOXP2 on the behaviour of the mouse. Nevertheless, further research on the genetics of language impairment could be extremely fruitful. If the idea of a special linguistic competence is correct, many genes must be involved. Only by finding and analysing these genes, as developmental genes are being analysed, can we ever understand the nature of 'universal grammar'. It may also help to answer the question, 'what were these genes doing before they acquired their present role?'

7.7.3 The evolution of language

Until recently, linguists have been reluctant to discuss the evolution of language, partly because they have found it hard to visualize intermediates between having no linguistic competence, and having 'universal grammar', and partly because there is in any case nothing corresponding to a fossil record whereby intermediates can be identified. The first break from this position was made by Pinker and Bloom (1990), who argued that, since any improvement in the ability to communicate would be adaptive, the obvious hypothesis is that linguistic competence, like other complex but adaptive organs, has evolved by natural selection. Bickerton (1990) proposed that the process occurred in two major steps—from ape signals to 'protolanguage', and from protolanguage to modern language.

What is protolanguage? Consider the following utterances, from two sources:

1.	2.
Big train; Red book.	Drink red; Comb black.
Walk street; Go store.	Go on; Look out.
Put book; Hit ball.	Tickle Washoe; Open blanket.

The first were spoken by children at the two-word stage; the second are by Washoe, a chimpanzee raised and taught sign language by humans (Gardner and Gardner 1974). The lists are formally similar. They illustrate, among other things, the attribution of qualities to objects, and the relation of 'agents' (actors) and 'patients' (objects of an action) to actions. Similar linguistic properties are found in two other contexts. One are the languages (called 'pidgins') used by adults with no common language, who must nevertheless work together: slaves in island colonies are an example. The final context in which protolanguage occurs is in the speech of adults (such as Genie, described above) who had no exposure to language as children. Although the examples given above are of two-word utterances, longer utterances occur, but they lack many features of modern language described below.

Bickerton's suggestion that protolanguage represents an earlier intermediate stage in language evolution seems entirely plausible: we discuss in a moment how it might have evolved from ape signalling. Unfortunately, he then suggests that the origin of modern language from protolanguage must have occurred by a 'macromutation'. Jackendoff (2002) proposes a scenario whereby protolanguage could have evolved from ape language, and could then have evolved, by a number of steps, into modern language (Fig. 7.2). We discuss his proposed steps in turn.

1. *Use of symbols in a non-situation-specific fashion* Monkeys and apes use symbols in a specific context. The Vervet Monkey's 'Leopard' alarm is used to assert 'there is a Leopard present', but cannot be used, for example, to ask if anyone has seen a Leopard recently, or to tell a juvenile that Leopards have spots. In contrast, a human baby may use the word 'kitty' to draw attention to a cat, to ask where the cat is, to call the cat, and so on. In captivity, apes learn to use the same symbol in different contexts, but no wild ape or monkey does so.

2. *An open class of symbols: 'phonemes'* The vocabulary of humans is numbered in thousands: in apes and monkeys in tens, whether wild or in captivity. Humans, hearing speech, first 'digitalise' the sound: that is, they break it up into a sequence of discrete sounds belonging to one of some 40 or so kinds (the kinds, of course, vary with the language). If sounds were not first classified in this way, it would be impossible to learn and remember a vocabulary even of one thousand words.

3. *Stringing symbols together: word order conveys meaning* The conversion of a finite vocabulary into a system able to express an indefinitely large number of meanings depends on the ability to string words together. Rules are needed to interpret multi-word utterances, even short ones. Two kinds of rule are possible. The first uses word order to convey meaning: the second introduces special grammatical symbols.

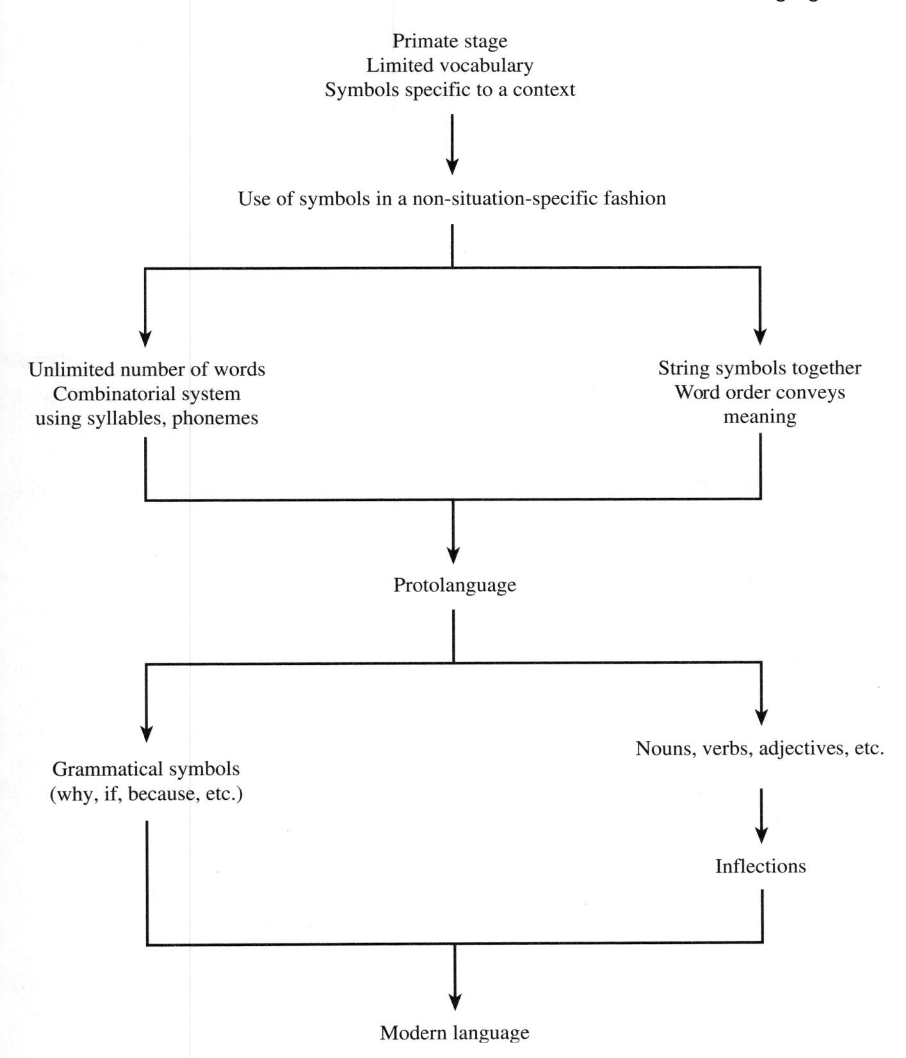

Fig. 7.2 Stages in the evolution of human language (after Jackendoff 2002, slightly simplified).

Protolanguage uses only the first of these methods, in a very simple form. For example, it uses the rule 'agent first; patient second'. Thus, 'tree Fred hit' means that the tree hit Fred, and 'Fred tree hit' that Fred hit the tree. The ability to use such simple rules is found in captive chimps that have been taught language (Savage-Rumbaugh *et al.* 1998). Captive dolphins can also learn such rules (Herman *et al.* 1984); for example, they can learn to distinguish 'hoop fetch pipe' (fetch the hoop to the pipe) and 'pipe fetch hoop' (fetch the pipe to the hoop). However, there is no convincing evidence for the use of such rules in wild animals

A combination of steps 1–3 converts primate signalling into protolanguage. A crucial next step is the acquisition of phrase structure. Compare the sentences:

> Boy ate apple.
>
> The boy in the T-shirt ate the apple his mother bought for him.

In the second sentence, the single noun 'Boy' has been replaced by the noun phrase 'The boy in the T-shirt', and the noun 'apple' by the noun phrase 'the apple his mother bought for him'. The meaning of the sentence now depends on phrase order. But it is hard for the hearer to recognize the phrases and to deduce their order without additional help. This can come either from special grammatical terms (e.g. what, and, but, the, if, which, because) which have no meaning on their own but which help to reveal the structure and sense of the sentence; or by 'inflections' (e.g. past and present tense of verbs, singular and plural). Phrase structure, plus these additional aids, provide the basis for modern language.

Jackendoff also suggests that it is possible to recognize, in modern language, 'fossils' of earlier stages, just as in morphological evolution the bones of the inner ear of mammals are a surviving relict of the jaw articulation of fish. We will give one example. We have seen that the rule 'agent first' is an important part of protolanguage, enabling the hearer to tell who did the hitting, and who was hit. The rule does not always hold in modern language—for example, in 'Fred was hit by the tree'. But the rule survives in modern language as a 'default principle', used to identify the agent in the absence of other clues. Thus the two sentences, 'the cat ate the mouse' and 'the mouse ate the cat', have no specific word forms that indicate who ate whom. We infer the 'agent' from the word order, as would be done by a speaker at the protolanguage stage of evolution.

Thanks to Pinker and Bloom, to Bickerton, and to Jackendoff, we now have a convincing picture of how language could have evolved by natural selection, through a series of functional intermediates. We believe that further progress will depend on genetic analysis of people with innate deficiencies in linguistic competence. It will not be easy, and will require collaboration between two professions that have hardly been on speaking terms for a century: but it will be worth it.

Glossary of scientific names

African Elephant	*Loxodonta africana*
African Grey Parrot	*Psiattacus erithacus*
African Swallowtail	*Papilio dardanus*
African Wild Dog	*Lycaon pictus*
American Goldfinch	*Carduelis tristis*
Arabian Babbler	*Turdoides squamiceps*
Atlantic Gannet	*Morus bassanus (Sula bassana)*
Australian Magpie	*Gymnorhina tibicen*
Bald Eagle	*Haliaeetus leucocephalus*
Bearded Tit	*Panurus biarmicus*
Belding's Ground Squirrel	*Spermophilus beldingi*
Big-clawed Snapping Shrimp	*Alpheus heterochaelis*
Bighorn Sheep	*Ovis canadensis*
Black Stork	*Ciconia nigra*
Blue-footed Booby	*Morus (Sula) nebouxii*
Bluehead Wrasse	*Thalassoma bifasciatum*
Bluethroat	*Luscinia svecica*
Brown-headed Cowbird	*Molothrus ater*
Campbell's Monkey	*Cercopithecus campbelli*
Canary	*Serinus canaria*
Caracal	*Caracal caracal*
Chacma Baboon	*Papio ursinus*
Chaffinch	*Fringilla coelebs*
Cheetah	*Acinonyx jubatus*
Chiffchaff	*Phylloscopus collybita*
Collared Flycatcher	*Ficedula albicollis*
Common Goldeneye	*Bucephala clangula*
Common Redpoll	*Carduelis flammea*
Common Toad	*Bufo bufo*
Crested Auklet	*Aethia cristatella*
Crested Tit	*Parus cristatus*
Crowned Eagle	*Stephanoaetus coronatus*
Diana Monkey	*Cercopithecus diana*
Double-crested Cormorant	*Phalacrocorax auritus*

Early Spider Orchid	*Ophrys sphegoides*
Earwig	*Forficula auricularia*
Elephant Seal	*Mirounga leonina*
Emperor Penguin	*Aptenodytes forsteri*
Eurasian Nuthatch	*Sitta europaea*
Eurasian Siskin	*Carduelis spinus*
European Robin	*Erithacus rubecula*
Fallow Deer	*Dama dama*
Funnel-web Spider	*Agelenopsis aperta*
Galah	*Cacatua roseicapilla*
Giant Deer	*Megaloceros giganteus*
Gladiator Frog	*Hyla rosenbergi*
Golden Jackal	*Canis aureus*
Golden-mantled Ground Squirrel	*Spermophilus lateralis*
Great Crested Grebe	*Podiceps cristatus*
Great Tit	*Parus major*
Greater Flamingo	*Phoenicopterus ruber*
Greenfinch	*Carduelis chloris*
Guppy	*Poecilia reticulata*
Harris's Sparrow	*Zonotrichia querula*
Hen Harrier	*Circus cyaneus*
Honey Bee	*Apis mellifera*
honeycreeper ('Amakihi)	*Loxops virens*
Honeyguide	*Indicator indicator*
Horned (Slavonian) Grebe	*Podiceps auritus*
House Finch	*Carpodacus mexicanus*
House Sparrow	*Passer domesticus*
Javanese Munia	*Lonchura leucogastroides*
Killdeer	*Charadrius vociferus*
Killer Whale	*Orcinus orca*
Leopard	*Panthera pardus*
Lesser Flamingo	*Phoenicopterus minor*
Lesser Spotted Woodpecker	*Dendrocopos minor*
Long-tailed Grassfinch	*Poephila acuticauda*
Lyrebird	*Menura novaehollandiae*
Magpie	*Pica pica*
Malachite Sunbird	*Nectarinia johnstoni*
Mallard	*Anas platyrhynchos*
Mandarin	*Aix galericulata*
Mara	*Dolichotis pagagonium*
Marsh Tit	*Parus palustris*
Marsh Warbler	*Acrocephalus palustris*
Martial Eagle	*Polemaetus bellicosus*
Mule Deer	*Odocileus hemionus*

Musk Duck	*Biziura lobata*
Musk Ox	*Ovibos moschatus*
Naked Mole Rat	*Heterocephalus glaber*
Narwhal	*Monodon monoceros*
Nightingale	*Luscinia megarhynchos*
Northern Fulmar	*Fulmarus glacialis*
Orangutan	*Pongo pygmaeus*
Oribi	*Ourebia ourebi*
Osprey	*Pandion haliaetus*
Otter	*Lutra lutra*
Oystercatcher	*Haematopus ostralegus*
Painted Redstart	*Myioborus pictus*
Peacock (Common Peafowl)	*Pavo cristatus*
Peewit (or Lapwing)	*Vanellus vanellus*
Peregrine Falcon	*Falco peregrinus*
Pied Wagtail	*Motacilla alba yarrellii*
Pigtail Macaque	*Macaca nemestrina*
Purple Finch	*Carpodacus purpureus*
Red Deer	*Cervus elephus*
Red Howler Monkey	*Alouatta seniculus*
Red-winged Blackbird	*Agelaius phoeniceus*
Rhesus Macaque	*Macaca mulatta*
Roe Deer	*Capreolus europaeus*
Ruddy Turnstone	*Arenaria interpres*
Ruff	*Philomachus pugnax*
Rufous-sided Towhee	*Pipilo erythrophthalmus*
Sage Grouse	*Centrocercus urophasianus*
Serval	*Leptailurus serval*
Shore Crab	*Carcinus maenas*
Silver-backed Jackal	*Canis mesomelas*
Silvereye	*Zosterops lateralis*
Silver-washed Fritillary	*Argynnis paphia*
Sparrowhawk	*Accipiter nisus*
Squirrel Monkey	*Saimiri sciureus*
Stag Beetle	*Lucanus cervus*
Stalk-eyed Fly	*Cyrtodiopsis dalmanni*
Starling	*Sturnus vulgaris*
Stellar's Jay	*Cyanocitta stelleri*
Superb Starling	*Lamprotornis (Spreo) superbus*
Swallow	*Hirundo rustica*
Thomson's Gazelle	*Gazella thomsonii*
Three-spined Stickleback	*Gasterosteus aculeatus*
Tiger	*Panthera tigris*
Tungara Frog	*Physalaemus pustulosus*

Varied Thrush	*Ixoreus naevius*
Vervet Monkey	*Cercopithecus aethiops*
Waterbuck	*Kobus ellipsiprymnus*
Western Grebe	*Aechmophorus occidentalis*
White Stork	*Ciconia ciconia*
White Wagtail	*Motacilla alba*
White-browed Robin Chat	*Cossypha heuglini*
White-crowned Sparrow	*Zonotrichia leucophrys*
Willow Tit	*Parus montanus*
Wood Duck	*Aix sponsa*
Zebra Finch	*Poephila (Taeniopygia) guttata*

References

Adams, E. S. and Caldwell, R. L. (1990). Deceptive communication in asymmetric fights of the stomatopod crustacean *Gonodactylus bredini*. *Animal Behaviour*, 39, 706–17.

Adams, E. S. and Mesterton-Gibbons, M. (1995). The cost of threat displays and the stability of deceptive communication. *Journal of Theoretical Biology*, 175, 405–21.

Andersson, M. (1980). Why are there so many threat displays? *Journal of Theoretical Biology*, 86, 773–81.

Andersson, M. (1986). Evolution of condition-dependent sex ornaments and mating preferences: sexual selection based on viability differences. *Evolution*, 40, 804–16.

Andrew, R. J. (1956). Some remarks on behaviour in conflict situations, with special reference to *Emberiza* spp. *British Journal of Animal Behaviour*, 4, 41–5.

Andrew, R. J. (1961). The displays given by passerines in courtship and reproductive fighting: a review. *Ibis*, 103, 315–48, 549–79.

Andrew, R. J. (1963). The origin and evolution of the calls and facial expressions of the primates. *Behaviour*, 20, 1–109.

Andrew, R. J. (1972). The information potentially available in mammal displays. In *Non-verbal communication* (ed. R. A. Hinde), pp. 179–206. Cambridge University Press, Cambridge.

Aubin, T., Jouventin, P., and Hildebrand, C. (2000). Penguins use the two-voice system to recognize each other. *Proceedings of the Royal Society of London B*, 267, 1081–7.

Backwell, P. R. Y., Christy, J. H., Telford, S. R., Jennions, M. D., and Passmore, N. I. (2000). Dishonest signalling in a fiddler crab. *Proceedings of the Royal Society of London B*, 267, 719–24.

Baenninger, R. (1997). On yawning and its functions. *Psychonomic Bulletin and Review*, 4, 198–297.

Baerends, G. P. (1975). An evaluation of the conflict hypothesis as an explanatory principle for the evolution of displays. In *Function and evolution of behaviour* (eds G. P. Baerends, C. Beer, and A. Manning), pp. 187–228. Clarendon Press, Oxford.

Baerends, G. P. and Baerends van Roon, J. M. (1950). An introduction to the study of the ethology of cichlid fishes. *Behaviour Supplement*, 1, 1–243.

Bakker, T. C. M., Kunzler, R., and Mazzi, D. (1999). Sexual selection—condition-related mate choice in sticklebacks. *Nature*, 401, 234.

Baptista, L. F. and Morton, M. L. (1981). Interspecific song acquisition by a White-crowned Sparrow. *Auk*, 98, 383–5.

Barber, I., Arnott, S. A., Braithwaite, V. A., Andrew, J., and Huntingford, F. A. (2001). Indirect fitness consequences of mate choice in sticklebacks: offspring of brighter males grow slowly but resist parasitic infections. *Proceedings of the Royal Society of London B*, 268, 71–6.

Basolo, A. L. (1990). Female preference predates the evolution of the sword in swordtail fish. *Science*, 250, 808–10.

Basolo, A. L. (1995). Phylogenetic evidence for the role of pre-existing bias in sexual selection. *Proceedings of the Royal Society of London B*, 259, 307–11.

Basolo, A. L. (1998). Evolutionary change in a receiver bias: a comparison of female preference functions. *Proceedings of the Royal Society of London B*, 265, 2223–8.

Basolo, A. L. (2002). Female discrimination against sworded males in a poeciliid fish. *Animal Behaviour*, 63, 463–8.

Belthoff, J. R., Dufty, A. M., and Gauthraux, S. A. (1994). Plumage variation, plasma steroids and social dominance in male House Finches. *Condor*, 96, 614–25.

Berglund, A. and Rosenqvist, G. (2001). Male pipefish prefer ornamented females. *Animal Behaviour*, 61, 345–50.

Berglund, A., Bisazza, A., and Pilastro, A. (1996). Armaments and ornaments: an evolutionary explanation of traits of dual utility. *Biological Journal of the Linnean Society*, 58, 85–399.

Bergstrom, C. T. and Lachmann, M. (1997). Signalling among relatives. I. Is costly signalling *too* costly? *Philosophical Trnsactions of the Royal Society of London B*, 352, 609–17.

Bergstrom, C. T. and Lachmann, M. (1998). Signalling among relatives. III. Talk is cheap. *Proceedings of the National Academy of Sciences of the United States of America*, 95, 5100–5.

Bickerton, D. (1990). *Language and species*. Chicago University Press, Chicago.

Birkhead, T. R. and Møller, A. P. (1998). *Sperm competition and sexual selection*. Academic Press, London.

Bishop, D. V. M. (2002). Putting language genes in perspective. *Trends in Genetics*, 18, 57–9.

Bishop, D. T., Cannings, C., and Maynard Smith, J. (1978). The war of attrition with random rewards. *Journal of Theoretical Biology*, 74, 377–88.

Bjorksten, T., David, P., Pomiankowski, A., and Fowler, K. (2000). Fluctuating asymmetry of sexual and nonsexual traits in stalk-eyed flies: a poor indicator of developmental stress and genetic quality. *Journal of Evolutionary Biology*, 13, 89–97.

Boesch, C. (1991). Symbolic communication in wild chimpanzees? *Human Evolution*, 6, 81–90.

Boesch, C. (1996). The emergence of cultures among wild chimpanzees. In *Evolution of social behaviour patterns in primates and man* (eds. W. G. Runciman, J. Maynard Smith, and R. I. M. Dunbar), pp. 251–68. Oxford University Press, Oxford.

Boyce, M. S. (1990). The Red Queen visits Sage Grouse leks. *American Zoologist*, 30, 263–70.

Bradbury, J. W. and Vehrencamp, S. L. (1998). *Principles of animal communication*. Sinauer, Sunderland, MA.

Brashares, J. S. and Arcese, P. (1999). Scent marking in a territorial antelope: II. The economics of marking with faeces. *Animal Behaviour*, 57, 11–17.

Breuker, C. J. and Brakefield, P. M. (2002). Female choice depends on size but not symmetry of dorsal eyespots in the butterfly *Bicyclus anynana*. *Proceedings of the Royal Society of London B*, 269, 1233–9.

Brown, J. (1975). *The Evolution of behaviour*. W. W. Norton, New York.

Brown, C. R., Brown, M. B., and Shaffer, M. L. (1991). Food sharing signals among socially foraging Cliff Swallows. *Animal Behaviour*, 42, 551–64.

Burgess, J. W. (1976). Social spiders. *Scientific American*, March, 100–6.

Burley, N. T. (1981). Sex-ratio manipulation and selection for attractiveness. *Science*, 211, 721–2.

Burley, N. T. (1988). Wild Zebra Finches have band-color preferences. *Animal Behaviour*, 36, 1235–7.

Burley, N. T., Enstrom, D. A., and Chitwood, L. (1994). Extra-pair relations in Zebra Finches: differential male success results from female tactics. *Animal Behaviour*, 48, 1031–41.

Burley, N. T. and Symanski, R. (1998). 'A taste for the beautiful'. Latent aesthetic mate preferences for white crests in two species of Australian grassfinches. *American Naturalist*, 152, 792–802.

Byrne, R. W. and Whiten, A. (1988). *Machiavellian intelligence: social expertise and the evolution of the intellect in monkeys, apes and humans*. Clarendon Press, Oxford.

Cardoso, M. Z. (1997). Testing chemical defence based on pyrrolizidine alkaloids. *Animal Behaviour*, 54, 985–91.

Caro, T. M. (1986). The functions of stotting in Thomson's Gazelles: some tests of the predictions. *Animal Behaviour*, 34, 663–84.

Carranza, J. and Mateos-Quesada, P. (2001). Habitat modification when scent-marking: shrub clearance by Roe Deer bucks. *Oecologia*, 126, 231–8.

Caryl, P. (1979). Communication by agonistic displays: what can game theory contribute to ethology? *Behaviour*, 68, 136–69.

Catchpole, C. K. and Slater, P. J. B. (1995). *Bird song: biological themes and variations*. Cambridge University Press, Cambridge.

Chanin, P. (1985). *The natural history of otters*. Croom Helm, London.

Cheney, D. L. and Seyfarth, R. M. (1982). How Vervet Monkeys perceive their grunts: field playback experiments. *Animal Behaviour*, 30, 739–51.

Cheney, D. L. and Seyfarth, R. M. (1988). Assessment of meaning and the detection of unreliable signals by Vervet Monkeys. *Animal Behaviour*, 36, 477–86.

Cheney, D. L. and Seyfarth, R. M. (1990). *How monkeys see the world*. Chicago University Press, Chicago.

Cheney, D. L. and Seyfarth, R. M. (1996). Function and intention in the calls of non-human primates. In *Evolution of social behaviour patterns in primates and man* (eds W. G. Runciman, J. Maynard Smith, and R. I. M. Dunbar), pp. 59–76. Oxford University Press, Oxford.

Cheney, D. L. and Seyfarth, R. M. (1997). Reconciliatory grunts by dominant females influence victim's behaviour. *Animal Behaviour*, 54, 409–18.

Cheney, D. L., Seyfarth, R. M., and Silk, J. B. (1995). The role of grunts in reconciling opponents and facilitating interactions among adult female baboons. *Animal Behaviour*, 50, 249–57.

Chomsky, N. T. (1965). *Aspects of the theory of syntax*. MIT press, Cambridge, MA.

Clutton-Brock, T. H. (1982). The functions of antlers. *Behaviour*, 79, 108–25.

Clutton-Brock, T. H. and Albon, S. D. (1979). The roaring of Red Deer and the evolution of honest advertisement. *Behaviour*, 69, 145–70.

Clutton-Brock, T. H. and Parker, G. A. (1995). Punishment in animal societies. *Nature*, 373, 209–16.

Clutton-Brock, T. H., Guinness, F. E., and Albon, S. D. (1982). *Red Deer: behaviour and ecology of two sexes*. Edinburgh University Press, Edinburgh.

Coutlee, E. (1967). Agonistic behaviour in the American Goldfinch. *Wilson Bulletin*, 79, 89–109.

Cramp, S. (1983). *Birds of the western Palearctic*. Vol. III. Oxford University Press, Oxford.

Cramp, S. (1985). *Birds of the western Palearctic*. Vol. V. Oxford University Press, Oxford.

Cramp, S. and Perrins, C. M. (1993). *Birds of the western Palearctic*. Vol. VII. Oxford University Press, Oxford.

Cramp, S. and Perrins, C. M. (1994). *Birds of the western Palearctic*. Vol. VIII. Oxford University Press, Oxford.

Cramp, S. and Simmons, K. E. L. (1977). *Birds of the western Palearctic*. Vol. I. Oxford University Press, Oxford.

Cullen, J. M. (1966). Reduction of ambiguity through ritualization. *Philosophical Transactions of the Royal Society of London B*, 251, 363–74.

Cumming, J. M. (1994). Sexual selection and the evolution of dance fly mating systems. *Canadian Entomologist*, 126, 907–20.

Cynx, J., Lewis, R., Tavel, B., and Tse, H. (1998). Amplitude regulation of vocalizations in noise by a songbird, *Taeniopygia guttata*. *Animal Behaviour*, 56, 107–13.

Dale, S. and Slagsvold, T. (1995). Female contests for nest sites and mates in the Pied Flycatcher *Ficedula hypoleuca*. *Ethology*, 99, 209–22.

Dane, B., Walcott, C., and Drury, W. H. (1959). The form and duration of the display actions of the Goldeneye (*Bucephala clangula*). *Behaviour*, 14, 265–81.

Darwin, C. (1871). *The descent of man and selection in relation to sex*. Murray, London.

Darwin, C. (1872). *The expression of the emotions in man and animals*. Murray, London.

David, P., Hingle, A., Greig, D., Rutherford, A., Pomiankowski, A., and Fowler, K. (1998). Male sexual ornament size but not asymmetry reflects condition in stalk-eyed flies. *Proceedings of the Royal Society of London B*, 265, 2211–16.

David, P., Bjorksten, T., Fowler, K., and Pomiankowski, A. (2000). A condition-dependent signalling of genetic variation in stalk-eyed flies. *Nature*, 406, 186–8.

Davies, N. B. (1981). Calling as an ownership convention on Pied Wagtail territories. *Animal Behaviour*, 29, 529–34.

Davies, N. B. and Halliday, T. R. (1978). Deep croaks and fighting assessment in toads, *Bufo bufo*. *Nature*, 274, 683–5.

Davies, N. B. and Houston, A. I. (1981). Owners and satellites: the economics of territory defence in the Pied Wagtail, *Motacilla alba*. *Journal of Animal Ecology*, 50, 157–80.

Davies, N. B. and Houston, A. I. (1983). Time allocation between territories and flocks and owner-satellite conflict in foraging Pied Wagtails, *Motacilla alba*. *Journal of Animal Ecology*, 52, 621–34.

Dawkins, M. S. (1986). *Unravelling animal behaviour*. Longman, Harlow.

Dawkins, M. S. and Guilford, T. (1994). Design of an intention signal in the bluehead wrasse, *Thalassoma bifasciatum*. *Proceedings of the Royal Society of London B*, 257, 123–8.

Dawkins, R. and Krebs, J. R. (1978). Animal signals: information or manipulation? In *Behavioural ecology: an evolutionary approach* (eds J. R. Krebs and N. B. Davies), pp. 282–309. Blackwell Scientific Publications, Oxford.

del Hoyo, J., Elliot, A., and Sargatal, J. (1992). *Handbook of the birds of the world*, vol. 1. Lynx Editions, Barcelona.

Dennett, D. C. (1987). *The intentional stance*. MIT/Bradford Books, Cambridge, MA.

Dilger, W. (1960). Agonistic and social behaviour of captive Redpolls. *Wilson Bulletin*, 72, 115–32.

Dimery, N. J., Alexander, R. McN., and Deyst, K. A. (1985). Mechanics of the ligamentum nuchae of some artiodactyls. *Journal of Zoology*, 206, 341–51.

Dowsett-Lemaire, F. (1979). The imitative range of the song of the Marsh Warbler *Acrocephalus palustris*, with special reference to imitations of African birds. *Ibis*, 121, 453–68.

Eberhard, W. G. (1977). Aggressive chemical mimicry by a bolas spider. *Science*, 198, 1173–5.

Eco, U. (1976). *A theory of semiotics*. Indiana University Press, Bloomington.

Eens, M., Pinxten, R., and Verheyen, R. F. (1992). Song learning in captive European Starlings, *Sturnus vulgaris*. *Animal Behaviour*, 44, 1131–43.

Ekman, P. (1992). Facial expressions of emotion: an old controversy and new findings. *Philosophical Transactions of the Royal Society of London B*, 335, 63–9.

Elgar, M. A. (1986). House Sparrows establish foraging flocks by giving chirrup calls if the resources are divisible. *Animal Behaviour*, 34, 169–74.

Endler, J. A. (1980). Natural selection on color pattern in *Poecilia reticulata*. *Evolution*, 34, 76–91.

Endler, J. A. (1983). Natural and sexual selection of color patterns in poeciliid fishes. *Environmental Biology of Fishes*, 9, 173–90.

Endler, J. A. (1991). Variation in the appearance of Guppy color patterns to guppies and their predators under different visual conditions. *Vision Research*, 31, 587–608.

Enquist, M. (1985). Communication during aggressive interactions with particular reference to variation in choice of behaviour. *Animal Behaviour*, 33, 1152–61.

Enquist, M. (2002). Spectacular phenomena and limits to rationality in genetic and cultural evolution. *Philosophical Transactions of the Royal Society of London B*, 357, 1585–94.

Enquist, M. and Jakobsson, S. (1986). Decision making and assessment in the fighting behaviour of *Nannacara anomala* (Cichlidae, Pisces). *Ethology*, 72, 143–53.

Enquist, M. and Leimar, O. (1983). Evolution of fighting behaviour: decision rules and assessment of relative strength. *Journal of Theoretical Biology*, 102, 387–410.

Enquist, M. and Leimar, O. (1987). Evolution of fighting behaviour: the effect of variation in resource value. *Journal of Theoretical Biology*, 127, 187–205.

Enquist, M., Plane, E., and Roed, J. (1985). Aggressive communication in Fulmars (*Fulmarus glacialis*) competing for food. *Animal Behaviour*, 33, 1007–20.

Enquist, M., Ghirlanda, S., and Hurd, P. L. (1998). Discrete conventional signalling of a continuous variable. *Animal Behaviour*, 56, 749–64.

Enquist, M., Leimar, O., Ljunberg, T., Mallner, Y., and Segerdahl, N. (1990). A test of the sequential assessment game: fighting in the cichlid fish *Nannacara anomala*. *Animal Behaviour*, 40, 1–14.

Evans, M. R. and Hatchwell, B. J. (1992). An experimental study of male adornment in the Scarlet-tufted Malachite Sunbird. I. The role of pectoral tufts in territorial defence. *Behavioural Ecology and Sociobiology*, 29, 413–19.

Fant, G. (1960). *Acoustic theory of speech production*. Mouton, The Hague.

Farrell, J. and Rabin, M. (1996). Cheap talk. *Journal of Economic Perspectives*, 10, 110–18.

Fehr, E. and Gachter, S. (2002). Altruistic punishment in humans. *Nature*, 415, 137–40.

Fisher, R. A. (1930). *The genetical theory of natural selection*. Clarendon Press, Oxford.

Fitch, W. T. (1997). Vocal tract length and formant frequency dispersion correlate with body size in Rhesus Macaques. *Journal of the Acoustical Society of America*, 102, 1213–22.

Fitch, W. T. and Reby, D. (2001). The descended larynx is not uniquely human. *Proceedings of the Royal Society of London B*, 268, 1669–75.

Fitze, P. S. and Richner, H. (2002). Differential effects of a parasite on ornamental structures based on melanins and carotenoids. *Behavioural Ecology*, 13, 401–7.

FitzGibbon, C. D. and Fanshawe, J. H. (1988). Stotting in Thomson's Gazelles: an honest signal of condition. *Behavioural Ecology and Sociobiology*, 23, 69–74.

Fitzpatrick, S. (1998). Colour schemes for birds: structural coloration and signals of quality in feathers. *Annales Zoologicae Fennica*, 35, 67–77.

Fleishman, L. J. (1992). The influence of the sensory system and the environment on motion patters in the visual displays of anoline lizards and other vertebrates. *American Naturalist*, 139, S36–61.

Funk, D. H. and Tallamy, D. W. (2000). Courtship role reversal and deceptive signals in the Long-tailed Dance Fly, *Rhamphomyia longicauda*. *Animal Behaviour*, 59, 411–21.

Galdikas, B. M. F. (1979). Orangutan adaptations at Tanjung Putting Reserve: mating and ecology. In *The great apes* (eds D. A. Hamburger and E. R. McCoun), pp. 79–98. Benjamin Cummings, Menlo Park.

Galdikas, B. M. F. (1995). *Reflections of Eden: my life with the orangutans of Borneo*. Victor Gollancz, London.

Galef, B. G. and Giraldeau, L.-A. (2001). Social influence on foraging in vertebrates: causal mechanisms and adaptive functions. *Animal Behaviour*, 61, 3–15.

Gamboa, G. J., Grudzien, T. A., Espelie, K. E., and Bura, E. A. (1996). Kin recognition pheromones in social wasps: combining chemical and behavioural evidence. *Animal Behaviour*, 51, 625–9.

Gardner, B. T. and Gardner, A. R. (1974). Comparing the early utterances of child and chimpanzee. In *Minnesota symposium on child psychology* (ed. A. Pick), vol. 8, pp. 3–23. University of Minnesota Press, Minneapolis.

Geist, V. (1966). The evolution of horn-like organs. *Behaviour*, 27, 175–213.

Ghirlanda, S. (2002). Intensity generalization: physiology and modelling of a neglected topic. *Journal of Theoretical Biology*, 214, 389–404.

Gibson, R. M. (1990). Relationships among blood parasites, mating success and phenotypic cues in male Sage Grouse *Centrocersus urophasianus*. *American Zoologist*, 30, 271–8.

Gillies, M. T. (1980). The role of carbon dioxide in host-finding by mosquitoes (Diptera: Culicidae): a review. *Bulletin of Entomological Research*: 70, 525–32.

Gintis, H. (2000). Strong reciprocity and human society. *Journal of Theoretical Biology*, 206, 169–79.

Glanville, E. V. (1992). Cooperative fishing by Double-crested Cormorants, *Phalacrocorax auritus*. *The Canadian Field-Naturalist*, 106, 522–3.

Godfray, H. C. J. (1991). Signalling of need by offspring to their parents. *Nature*, 352, 328–30.

Godfray, H. C. J. (1995). Signalling of need between parents and young: parent–offspring conflict and sibling rivalry. *American Naturalist*, 146, 1–24.

Godfray, H. C. J. and Johnstone, R. A. (2000). Begging and bleating: the evolution of parent–offspring signalling. *Proceedings of the Royal Society of London B*, 355, 1581–91.

Godfray, H. C. J. and Parker, G. A. (1992). Sibling competition, parent–offspring conflict and clutch size. *Animal Behaviour*, 43, 473–90.

Goodall, J. (1971). *In the shadow of man*. Collins, London.

Goodall, J. (1986). *The Chimpanzees of the Gombe: patterns of behaviour*. Harvard University Press, Cambridge, MA.

Gopnik, M. (1990). Feature-blind grammar and dysphasia. *Nature*, 344, 715.

Gosling, L. M. (1982). A reassessment of the function of scent marking in territories. *Zeitschrift fur Tierpsychologie*, 60, 89–118.

Gosling, L. M. and McKay, H. V. (1990). Competitor assessment by scent-matching: an experimental test. *Behavioural Ecology and Sociobiology*, 26, 415–20.

Gosling, L. M. and Roberts, S. C. (2001). Scent-marking by male mammals: cheat-proof signals to competitors and mates. *Advances in the Study of Behaviour*, 30, 169–217.

Gould, S. J. (1998). *Leonardo's mountain of clams and the diet of worms*. Cape, London.

Gouzoules, H. and Gouzoules, S. (1989). Design features and developmental modification of Pigtail Macaque, *Macaca nemestrina*, agonistic screams. *Animal Behaviour*, 37, 383–401.

Gouzoules, S., Gouzoules, H., and Marler, P. (1984). Rhesus monkey (*Macaca mulatta*) screams: representational signalling in the recruitment of agonistic aid. *Animal Behaviour*, 32, 182–93.

Grafen, A. (1987). The logic of divisively asymmetric contests: respect for ownership and the desperado effect. *Animal Behaviour*, 35, 462–7.

Grafen, A. (1990*a*). Biological signals as handicaps. *Journal of Theoretical Biology*, 144, 517–546.

Grafen, A. (1990*b*). Sexual selection unhandicapped by the Fisher process. *Journal of Theoretical Biology*, 144, 475–518.

Green, D. J. and Krebs, E. A. (1995). Courtship feeding in Ospreys *Pandion haliaetus*—a criterion for mate assessment. *Ibis*, 137, 35–43.

Gregory, R. and Hopkins, P. (1974). Pupils of a talking parrot. *Nature*, 252, 637–8.

Grether, G. F. (2000). Carotenoid limitation and mate preference evolution: a test of the indicator hypothesis in guppies (*Poecilia reticulata*). *Evolution*, 54, 1712–24.

Grosberg, R. K. and Quinn, J. F. (1986). The genetic control and consequences of kin recognition by the larvae of a colonial marine invertebrate. *Nature*, 322, 457–9.

Guilford, T. and Dawkins, M. S. (1991). Receiver psychology and the evolution of animal signals. *Animal Behaviour*, 42, 1–14.

Guilford, T. and Dawkins, M. S. (1995). What are conventional signals? *Animal Behaviour*, 49, 1689–95.

Guinet, C. (1992). Predation behaviour of Killer Whales (*Orcinus orca*) around Crozet islands. *Canadian Journal of Zoology*, 70, 1656–67.

Gustaffson, L., Qvarnstrom, A., and Sheldon, B. C. (1995). Trade-offs between life-history traits and a secondary sexual character in male Collared Flycatchers. *Nature*, 375, 311–13.

Haftorn, S. (2000). Contexts and possible functions of alarm calling in the Willow Tit, *Parus montanus*; The principle of 'better safe than sorry'. *Behaviour*, 137, 437–49.

Hamilton, W. D. (1964). The genetical evolution of social behaviour, I and II. *Journal of Theoretical Biology*, 7, 1–16, 17–32.

Hamilton, W. D. (1979). Wingless fighting males in fig wasps and other insects. In *Sexual selection and reproductive competition in insects* (eds M. S. Blum and N. A. Blum), pp. 167–220. Academic Press, London.

Hamilton, W. D. and Zuk, M. (1982). Heritable true fitness and bright birds: a role for parasites. *Science*, 218, 384–7.

Hansen, A. J. (1986). Foraging behavior of Bald Eagles: a test of evolutionary game theory. *Ecology*, 67, 787–97.

Hansen, A. J. and Rohwer, S. (1986). Coverable badges and resource defence in birds. *Animal Behaviour*, 34, 69–76.

Hanson, H. M. (1959). Effects of discrimination training on stimulus generalization. *Journal of Experimental Psychology*, 58, 321–33.

Haskell, D. (1994). Experimental evidence that nestling begging behaviour incurs a cost due to nest predation. *Proceedings of the Royal Society of London B*, 257, 161–4.

Hasson, O. (1994). Cheating signals. *Journal of Theoretical Biology*, 167, 223–38.

Hasson, O. (1991). Pursuit-deterrant signals: communication between prey and predator. *Trends in Evolution and Ecology*, 6, 325–9.

Hasson, O., Cohen, D., and Shmida, A. (1992). Providing or hiding information: on the evolution of amplifiers and attenuators of perceived quality differences. *Acta Biotheoretica*, 40, 269–83.

Hauber, M. E., Russo, S. A., and Sherman, P. W. (2001). A password for species recognition in a brood-parasitic bird. *Proceedings of the Royal Society of London B*, 268, 1041–8.

Hauser, M. D. (1988). How infant Vervet Monkeys learn to recognize starling alarm calls: the role of experience. *Behaviour*, 105, 187–201.

Hauser, M. D. (1996). *The evolution of communication*. MIT Press, Cambridge, MA.

Hauser, M. D. and Marler, P. (1993). Food-associated calls in rhesus macaques (*Macacca mulatta*): II. Costs and benefits of call production and suppression. *Behavioural Ecology*, 4, 206–12.

Herberholz, J. and Schmitz, B. (1998). Role of mechanosensory stimuli in intraspecific agonostic encounters of the Snapping Shrimp (*Alpheus heterochaelis*). *Biological Bulletin*, 195, 156–67.

Herman, L. M., Richards, D. G., and Wolz, J. P. (1984). Comprehension of sentences by Bottle-nosed Dolphins. *Cognition*, 16, 129–219.

Hess, E. H. (1965). Attitude and pupil size. *Scientific American*, 212, 46–54

Hill, G. E. (1992). Proximate basis of variation in carotenoid pigmentation in male House Finches. *Auk*, 109, 1–12.

Houde, A. E. and Torio, A. J. (1992). Effect of parasitic infection on male color pattern and female choice in guppies. *Behavioural Ecology*, 3, 346–51.

Hughes, M. (2000). Deception with honest signals: signal residuals and signal function in Snapping Shrimp. *Behavioural Ecology*, 11, 614–23.

Hughes, M., Hultsch, M., and Todt, D. (2002). Initiation and invention in song learning in Nightingales (*Luscinia megarhynchos*). *Ethology*, 108, 97–113.

Hunt, J. and Simmons, L. W. (1997). Patterns of fluctuating asymmetry in beetle horns: an experimental examination of the honest signalling hypothesis. *Behavioural Ecology and Sociobiology*, 41, 109–14.

Hunter, M. L. and Krebs, J. R. (1979). Geographic variation in the song of the Great Tit (*Parus major*) in relation to ecological factors. *Journal of Animal Ecology*, 48, 759–85.

Huntingford, F. and Turner, A. K. (1987). *Animal conflict*. Chapman & Hall, London.

Hurd, P. L. (1997). Cooperative signalling between opponents in fish fights. *Animal Behaviour*, 54, 1309–15.

Hurd, P. L. and Enquist, M. (1998). Conventional signalling in aggressive interactions: the importance of temporal structure. *Journal of Theoretical Biology*, 192, 197–211.

Hurd, P. L. and Enquist, M. (2001). Threat display in birds. *Canadian Journal of Zoology*, 79, 931–42.

Huxley, J. S. (1930). *Bird watching and bird behaviour*. Chatto & Windus, London.

Isack, H. A. and Reyer, H.-U. (1989). Honey guides and honey-gatherers: interspecific communication in a symbiotic relationship. *Science*, 243, 1343–6.

Jablonski, P. G. and Strausfeld, N. J. (2000). Exploitation of an ancient escape circuit by an avian predator: prey sensitivity to model predator display in the field. *Brain, Behaviour and Evolution*, 56, 94–106.

Jackendoff, R. (2002). *Foundations of language*. Oxford University Press, Oxford.

Jackson, W. M., Rohwer, S., and Winnegrad, R. L. (1988). Social status signalling is absent within age and sex classes of Harris' Sparrows. *Auk*, 105, 424–7.

Jarvi, T. and Bakken, M. (1984). The function of the variation in the breast stripe of the Great Tit *Parus major*. *Animal Behaviour*, 32, 590–6.

Jarvi, T., Walso, O., and Bakken, M. (1987). Status signalling by *Parus major*, an experiment in deception. *Ethology*, 76, 334–42.

Jenkins, H. M. and Harrison, R. H. (1960) Effect of discrimination training on auditory generalization. *Journal of Experimental Psychology*, 59, 246–53.

Jennings, D. J., Gammell, M. P., Carlin, C. M., and Hayden, T. J. (2002). Does lateral presentation of the palmate antlers during fights by Fallow Deer (*Dama dama L.*) signify dominance or submission? *Ethology*, 108, 389–401.

Jennions, M. D. (1998). The effect of leg band symmetry on female–male association in Zebra Finches. *Animal Behaviour*, 55, 61–7.

Johnsen, A., Fiske, P., Amundsen, T. Lifjeld, J. T., and Rohde, P. A. (2000). Colour bands, mate choice and paternity in the Bluethroat. *Animal Behaviour*, 59, 111–19.

Johnson, K., Dalton, R., and Burley, N. (1993). Preference of female American Goldfinches *Carduelis tristis* for natural and artificial male traits. *Behavioural Ecology*, 4, 138–43.

Johnson, L. L. and Boyce, M. S. (1991). Femle choice of males with low parasite loads in Sage Grouse. In *Bird–parasite interactions: ecology, evolution and behaviour* (ed. J. E. Loye and M. Zuk), pp. 377–88. Oxford University Press, Oxford.

Johnstone, R. A. (1994). Honest signalling, perceptual error and the evolution of 'all-or-nothing' displays. *Proceedings of the Royal Society of London B*, 256, 169–75.

Johnstone, R. A. (1997). The evolution of animal signals. In *Behavioural ecology* (eds J. R. Krebs and N. B. Davies), pp. 155–78. Oxford University Press, Oxford.

Johnstone, R. A. and Grafen, A. (1992). The continuous Sir Philip Sidney game: a simple model of biological signalling. *Journal of Theoretical Biology*, 156, 215–34.

Johnstone, R. A. and Norris, K. (1993). Badges of status and the cost of aggression. *Behavioural Ecology and Sociobiology*, 32, 127–34.

Jones, A. G., Walker, D., and Avise, J. C. (2001). Genetic evidence for extreme polyandry and extraordinary sex-role reversal in a pipefish. *Proceedings of the Royal Society of London B*, 268, 2531–2535.

Jones, I. L. and Hunter, F. M. (1999). Experimental evidence for mutual inter-and intrasexual selection favouring a Crested Auklet ornament. *Animal Behaviour*, 57, 521–8.

Kalmijn, A. J. (1982). Electric and magnetic-field detection in elasmobranch fishes. *Science*, 218, 916–18.

Kamil, A. C. (1978). Systematic foraging for nectar by Amakihi *Loxops virens*. *Journal of Comparative Physiology and Psychology*, 92, 388–96.

Kennedy, M., Spencer, H. G., and Gray, R. D. (1996). Hop, step and gape: do the social displays of the Pelecaniformes reflect phylogeny? *Animal Behaviour*, 51, 273–91.

Keys, G. C. and Rothstein, S. I. (1991). Benefits and costs of dominance and subordinance in white-crowned sparrows and the paradox of status signalling. *Animal Behaviour*, 42, 899–912.

Keyser, A. J. and Hill, G. E. (1999). Condition-dependent variation in the blue–ultraviolet coloration of a structurally based plumage ornament. *Proceedings of the Royal Society of London B*, 266, 771–7.

Kilner, R. M. (2001). A growth cost of begging in captive Canary chicks. *Proceedings of the National Academy of Sciences of the United States of America*, 98, 11394–8.

Kilner, R. M. and Johnstone, R. (1997). Begging the question: are offspring solicitation behaviours signals of need? *Trends in Evolution and Ecology*, 12, 11–15.

Kluge, A. G. (1981). The life history, social organisation, and parental behaviour of *Hyla rosenbergii* Boulenger, a nest-building gladiator frog. *Miscellaneous Publications of the Museum of Zooology, University of Michigan*, 160, 1–170.

Krams, I. (2001). Communication in Crested Tits and the risk of predation. *Animal Behaviour*, 61, 1065–8.

Krebs, J. R. (1982). Territorial defence in the Great Tit *Parus major*: do residents always win? *Behavioural Ecology and Sociobiology*, 22, 79–84.

Krebs, J. R. (1987). The evolution of animal signals. In *Mindwaves: thoughts on intelligence, identity and consciousness* (eds C. Blakemore and S. Greenfield), pp. 163–73. Blackwell, Oxford.

Krebs, J. R. and Dawkins, R. (1984). Animal signals: mind-reading and manipulation. In: *Behavioural ecology: an evolutionary approach* (eds J. R. Krebs and N. B. Davies), 2nd edn, pp. 380–402. Blackwell Scientific Publications, Oxford.

Lachmann, M., Bergstrom, C. T., and Szamado, S. (2000). The death of costly signalling? *Santa Fe Institute Working Paper Series* (00-12-072).

Lai, C. S., Fisher, S. E., Hurst, J. A., Vargha-Khaden, F., and Monaco, A. P. (2001). A forkhead-domain gene is mutated in a severe speech and language disorder. *Nature*, 413, 519–23.

LaPorte, J. (2002). Must signals handicap? *The Monist*, 85, 86–104.

Lass, N. J. and Brown, W. S. (1978). Correlational study of speakers' heights, weights, body surface areas, and speaking fundamental frequencies. *Journal of the Acoustical Society of America*, 63, 1218–20.

Ledyard, J. O. (1995). Public goods: a survey of experimental research. In *The handbook of experimental economics* (eds J. H. Kagel and A. E. Roth), pp. 111–94. Princeton University Press, Princeton.

Lemel, J. and Wallin, K. (1993). Status signalling, motivational condition and dominance—an experimental study in the Great Tit, *Parus major* L. *Animal Behaviour*, 45, 549–58.

Lieberman, P. (1984). *The biology and evolution of language*. Harvard University Press, Cambridge, MA.

Liker, A. and Barta, Z. (2001). Male badge size predicts dominance against females in House Sparrows. *Condor*, 103, 151–7.

Lingle, S. (1993). Escape gaits of White-tailed Deer, Mule Deer, and their hybrids—body configuration, biomechanics, and function. *Canadian Journal of Zoology*, 71, 708–24.

Lister, A. M. (1994). The evolution of the Giant Deer, *Megaloceros giganteus* (Blumenbach). *Journal of the Linnean Society of London*, 112, 65–100.

Little, A. C., Burt, D. M., Penton-Voak, I. S., and Perrett, D. I. (2001). Self-perceived attractiveness influences human female preferences for sexual dimorphism and symmetry in male faces. *Proceedings of the Royal Society of London B*, 268, 39–44.

Lorenz, K. (1939). Vergleichende Verhaltensforschung. *Zoologische Anzeiger, Supplement*, 12, 69–102.

Lorenz, K. (1966). *On aggression.* Methuen, London.

Lorenz, K. (1970). *Studies in animal and human behaviour*, vol. 1. Methuen, London.

MacDonald, D. W. (1984). *The encylopaedia of mammals*, vols 1 and 2. George Allen and Unwin, London.

MacDonald, D. W. (1985). The rodents IV: suborder hystricomorpha. In *Social odours in mammals* (eds R. E. Brown and D. W. MacDonald), pp. 480–506. Clarendon Press, Oxford.

Mackintosh, N. J. (1974). *The psychology of animal learning.* Academic Press, London.

Marchetti, K. (1993). Dark habitats and bright birds illustrate the role of the environment in species divergence. *Nature* 362, 149–52.

Marler, P. (1955). Characteristics of some alarm calls. *Nature*, 176, 6–8.

Marler, P. (1970). A comparative approach to vocal learning: song development in White-Crowned Sparrows. *Journal of Comparative Physiology and Psychology*, 71(Suppl.), 1–25.

Marler, P. and Evans, C. (1996). Birds calls: just emotional displays or something more? *Ibis*, 138, 26–33.

Marshall, A. J. (1950). The function of vocal mimicry in birds. *Emu*, 50, 5–16.

Martin, F. D. and Hengstebeck, M. F. (1981). Eye colour and aggression in juvenile Guppies, *Poecilia reticulata* Peters (Pisces: Poecillidae). *Animal Behaviour*, 29, 325–31.

Martin, S. (1970). The agonistic behaviour of Varied Thrushes (*Ixoreus naevius*) in winter assemblages. *Condor*, 72, 452–9.

Mateo, J. M. (2002). Kin recognition abilities and nepotism as a function of sociality. *Proceedings of the Royal Society of London B*, 269, 721–7.

Mather, K. (1953). Genetic control of stability in development. *Heredity*, 7, 297–336.

Matsuoka, S. (1980). Pseudo warning calls in titmice. *Tori*, 29, 87–90.

Matthysen, E. (1998). *The Nuthatches.* Poyser, London.

Maynard Smith, J. (1956). Fertility, mating behaviour and sexual selection in *Drosophila subobscura. Journal of Genetics*, 54, 261–79.

Maynard Smith, J. (1958). Sexual selection. In *A century of Darwin* (ed. S. A. Barnett), pp. 231–44. London, Heinemann.

Maynard Smith, J. (1976). Sexual selection and the handicap principle. *Journal of Theoretical Biology*, 57, 239–42.

Maynard Smith, J. (1982). *Evolution and the theory of games.* Cambridge University Press, Cambridge.

Maynard Smith, J. (1985). Sexual selection, handicaps and true fitness. *Journal of Theoretical Biology*, 115, 1–8.

Maynard Smith, J. (1991). Honest signalling: the Philip Sidney game. *Animal Behaviour*, 42, 1034–5.

Maynard Smith, J. and Brown, R. L. W. (1986). Competition and body size. *Theoretical Population Biology*, 30, 166–79.

Maynard Smith, J. and Harper, D. G. C. (1988). The evolution of aggression: can selection generate variability? *Philosophical Transactions of the Royal Society of London B*, 319, 557–70.

Maynard Smith, J. and Harper, D. G. C. (1995). Animal signals: models and terminology. *Journal of Theoretical Biology*, 177, 305–11.

Maynard Smith, J. and Parker, G. (1976). The logic of asymmetric contests. *Animal Behaviour*, 24, 159–75.

Maynard Smith, J. and Price, G. R. (1973). The logic of animal conflicts. *Nature, London*, 246, 15–18.

Maynard Smith, J. and Riechert, S. E. (1984). A conflicting-tendency model of spider agonistic behaviour: hybrid—pure population line comparisons. *Animal Behaviour*, 32, 564–78.

McCarty, J. P. (1996). The energetic cost of begging in nestling passerines. *Auk*, 113, 178–88.

McComb, K. E. (1991). Female choice for high roaring rates in Red Deer, *Cervus elephus. Animal Behaviour*, 41, 79–88.

McComb, K. E., Packer, C., and Pusey, A. (1994). Roaring and numerical assessment in contests between groups of female lions. *Animal Behaviour*, 47, 379–87.

McFarland, D. and Sibley, R. (1975). The behavioural final common path. *Philosophical Transactions of the Royal Society of London B*, 270, 265–93.

McGraw, K. J. and Hill, G. E. (2000). Carotenoid-based ornamentation and status signalling in the House Finch. *Behavioural Ecology*, 11, 520–7.

McMann, S. (2000). Effects of residence time on displays during territory establishment in a lizard. *Animal Behaviour*, 59, 513–22.

Merila, J., Sheldon, B. C., and Lindstrom, K. (1999). Plumage brightness in relation to haematozoan infections in the Greenfinch: bright males are a good bet. *Ecoscience*, 6, 12–18.

Metz, K. J. and Weatherhead, P. J. (1992). Seeing red—uncovering badges in Red-winged Blackbirds. *Animal Behaviour*, 43, 223–9.

Meyer, A., Morrisey, J., and Schartl, M. (1994). Molecular phylogeny of fishes of the genus *Xiphophorus* suggests repeated evolution of a sexually selected trait. *Nature*, 368, 539–41.

Møller, A. P. (1987*a*). Variation in badge size in male House Sparrows *Passer domesticus*: evidence for status signalling. *Animal Behaviour*, 35, 1637–44.

Møller, A. P. (1987*b*). Social control of deception among status signalling House Sparrows *Passer domesticus. Behavioural Ecology and Sociobiology*, 20, 307–11.

Møller, A. P. (1988). False alarm calls as a means of resource usurpation in the Great Tit *Parus major. Ethology*, 79, 25–30.

Møller, A. P. (1990). Sexual behaviour is related to badge size in the House Sparrow *Passer domesticus. Behavioural Ecology and Sociobiology*, 27, 23–9.

Møller, A. P. (1992). Female swallow preference for symmetrical male sexual ornaments. *Nature*, 357, 238–40.

Møller, A. P. (1993). Patterns of fluctuating asymmetry in sexual ornaments predict female choice. *Journal of Evolutionary Biology*, 6, 481–91.

Møller, A. P. and Thornhill, R. (1997). A meta-analysis of the heritability of developmental stability. *Journal of Evolutionary Biology*, 10, 1–16.

Møller, A. P. and Thornhill, R. (1998). Bilateral symmetry and sexual selection: a meta-analysis. *American Naturalist*, 151, 174–92.

Morris, D. J. (1956). The feather postures of birds and the problem of the origin of social signals. *Behaviour*, 9, 75–113.

Morris, D. J. (1957). 'Typical intensity' and its relation to the problem of ritualization. *Behaviour*, 11, 1–12.

Morton, E. S. (1975). Ecological sources of selection on avian sound. *American Naturalist*, 109, 17–34.

Munn, C. A. (1986*a*). Birds that 'cry wolf'. *Nature*, 319, 143–5.

Munn, C. A. (1986*b*). The deceptive use of alarm calls by sentinel species in mixed flocks of Neotropical birds. In *Deception: perspectives on human and non-human deceit* (eds R. W. Marshall and N. S. Thompson), pp. 169–75. State University of New York Press, Albany.

Nelson, J. B. (1978). *The Sulidae: Gannets and Boobies.* Oxford University Press, Oxford.

Noldeke, G. and Samuelson, L. (1999). How costly is the honest signalling of need? *Journal of Theoretical Biology*, 197, 527–39.

Norris, K. J. (1990). Female choice and the evolution of conspicuous plumage coloration of monogamous male Great Tits. *Behavioural Ecology and Sociobiology*, 26, 129–38.

Norris, K. J. (1993). Heritable variation in a plumage indicator of viability in male Great Tits *Parus major*. *Nature*, 362, 537–39.

Oehlert, B. (1958). Kampf und Paarbilding einiger Cichliden. *Zeitschrift fur Tierpsychologie*, 15, 141–74.

Ogilvie, M. and Ogilvie, C. (1986). *Flamingos.* Alan Sutton Press, Gloucester.

Osorio, D. and Ham, A. D. (2002). Spectral reflectance and directional properties of structural coloration in bird plumage. *Journal of Experimental Biology*, 205, 2017–27.

Owens, I. P. F. and Hartley, I. R. (1991). 'Trojan sparrows'; Evolutionary consequences of dishonest invasion for the badges-of-status model. *American Naturalist*, 138, 1187–205.

Owings, D. and Hennessey, D. (1984). The importance of variation in sciurid visual and vocal communication. In *Biology of ground dwelling squirrels* (eds J. O. Murie and G. R. Michener). University of Nebraska Press, Lincoln.

Parker, G. (1974). Assessment strategy and the evolution of fighting behaviour. *Journal of Theoretical Biology*, 47, 223–43.

Paton, D. and Caryl, P. (1986). Communication by agonistic displays: I. Variation in information content between samples. *Behaviour*, 98, 213–39.

Payne, R. B. and Payne, L. L. (1994). Song mimicry and species associations of West African indigo birds *Vidua* with Quail-finch *Ortygospiza atricollis*, Goldbreast *Amandava subflava* and Brown Twinspot *Clytospiza monteiri*. *Ibis*, 136, 291–304.

Penton-Voak, I. S., Perrett, D. I., Castles, D. L., Kobyashi, T., Burt, D. M., Murray, L. K., and Minamisawa, R. (1999). *Menstrual cycle alters face preference. Nature*, 399, 741–2.

Penton-Voak, I. S., Jones, B. C., Little, A. C., Baker, S., Tiddeman, B., Burt, D. M., and Perrett, D. I. (2001). Symmetry, sexual dimorphism in facial proportions and male facial atrractiveness. *Proceedings of the Royal Society of London B*, 268, 1617–23.

Pepperberg, I. M. (1999). *The Alex studies: cognitive and communicative abilities of Grey Parrots.* Harvard University Press, Cambridge, MA.

Perrett, D. I., Burt, D. M., Penton-Voak, I. S., Lee, K. J., Rowland, D. A., and Edwards, R. (1999). Symmetry and human facial attractiveness, *Evolution and Human Behaviour*, 20, 295–307.

Petrie, M., Halliday, T. R., and Sanders, C. (1991). Peahens prefer males with elaborate trains. *Animal Behaviour*, 41, 323–31.

Pickering, S. P. C. and Duverge, I. (1992). The influence of visual-stimuli provided by mirrors on the marching displays of Lesser Flamingos, *Phoeniconais minor*. *Animal Behaviour*, 43, 1048–50.

Pinker, S. and Bloom, P. (1990). Natural language and natural selection. *Behavioural and Brain Sciences*, 13, 707–26.

Plenge, M., Curio, E., and Witte, K. (2000). Sexual imprinting supports the evolution of novel male traits by transference of a preference for the colour red. *Behaviour*, 137, 741–58.

Pomiankowski, A. (1987). Sexual selection: the handicap principle does work—sometimes. *Proceedings of the Royal Society of London B*, 231, 123–45.

Pomiankowski, A. and Iwasa, Y. (1998). Runaway ornament diversity caused by Fisherian sexual selection. *Proceedings of the National Academy of Sciences USA*, 95, 5106–11.

Pond, C. M. (1978). Morphological aspects and the ecological and mechanical consequences of fat deposition in wild vertebrates. *Annual Review of Ecology and Systematics*, 9, 519–70.

Poole, J. H. (1989). Announcing intent: the aggressive state of musth in African Elephants. *Animal Behaviour*, 37, 140–52.

Popp, J. (1987). Risk and effectiveness in the use of agonistic displays by American Goldfinches. *Behaviour*, 103, 141–56.

Proctor, H. C. (1991). Courtship in the water mite, *Neumania papillator*: males capitalize on female adaptations for predation. *Animal Behaviour*, 42, 589–98.

Qvarnstrom, A. (1997). Experimentally increased badge size increases male competition and reduces male parental care in the Collared Flycatcher. *Proceedings of the Royal Society of London B*, 264, 1225–31.

Rand, M. S. (1990). Polymorphic sexual coloration in the lizard *Scoloporus undulatus erythrocheilus*. *American Midland Naturalist*, 124, 352–8.

Ratnieks, F. L. W. and Reeve, H. K. (1992). Conflict in single-queen hymenopteran societies: the structure of conflict and processes that reduce conflict in advanced eusocial societies. *Journal of Theoretical Biology*, 158, 33–65.

Reby, D., Hewison, M., Izquierdo, M., and Pepin, D. (2001). Red Deer (*Cervus elephus*) hinds discriminate between the roars of their current harem-holder stag and those of neighbouring stags. *Ethology*, 107, 951–9.

Reby, D. and McComb, K. (2003). Anatomical constraints generate honesty: acoustic cues to age and weight in the roars of Red Deer stags. *Animal Behaviour* 65, 317–29.

Redondo, T. and Castro, F. (1992). Signalling of nutritional need by magpie nestlings. *Ethology*, 92, 193–204.

Reeve, E. C. R. (1960). Some genetic tests on asymmetry of sternopleural chaeta number in *Drosophila*. *Genetical Research*, Cambridge, 1, 151–72.

Rendall, D., Seyfarth, R. M., Cheney, D. L., and Owren, M. J. (1999). The meaning and function of grunt variants in baboons. *Animal Behaviour*, 57, 583–92.

Reynolds, V. (1965). *Budongo: a forest and its Chimpanzees*. London, Methuen.

Rice, W. (1996). Sexually antagonistic male adaptation triggered by experimental arrest of female evolution. *Nature*, 381, 232–4.

Richards, D. G. (1981). Alerting and message components in songs of Rufous-sided Towhees. *Behaviour*, 76, 223–49.

Riechert, S. E. (1978). Games spiders play: behavioral variability in territorial disputes. *Behavioral Ecology and Sociobiology*, 3, 135–62.

Riechert, S. E. (1984). Games spiders play: III. Cues underlying context-associated changes in agonistic behaviour. *Animal Behaviour*, 32, 1–15.

Riechert, S. E. and Maynard Smith, J. (1989). Genetic analysis of two behavioural traits linked to fitness in the desert spider *Agelenopsis aperta*. *Animal Behaviour*, 37, 624–37.

Rodd, H. F., Hughes, K. A., Grether, G. F., and Baril, C. T. (2002). A possible non-sexual origin of mate preference: are male guppies mimicking fruit? *Proceedings of the Royal Society of London B*, 269, 475–81.

Rogers, L. J. and Kaplan, G. (1998). *Not only roars and rituals*. Allen and Unwin, St. Leonards, Australia.

Rohwer, S. (1977). Status signalling in Harris' Sparrows: some experiments in deception. *Behaviour*, 61, 107–29.

Rohwer, (1978). Reply to Shields on avian plumage variability. *Evolution*, 32, 670–3.

Rohwer, S. and Rohwer, F. C. (1978). Status signalling in Harris' Sparrows: experimental deceptions achieved. *Animal Behaviour*, 26, 1012–22.

Roper, T. J. (1986). Badges of status in avian societies. *New Scientist*, 109, 38–40.

Roper, T. J. (1990). Responses of domestic chicks to artificially coloured insect prey: effects of previous experience and background colour. *Animal Behaviour*, 39, 466–73.

Rowe, C. (2002). Sound improves visual discrimination in avian predators. *Proceedings of the Royal Society of London B*, 269, 1353–7.

Ryan, M. J. (1985). *The Túngara Frog: a study in sexual selection and communication*. University of Chicago Press, Chicago.

Ryan, M. J. (1990). Sexual selection, sensory systems and sensory exploitation. *Oxford Surveys in Evolutionary Biology*, 7, 157–95.

Ryan, M. J. (1997). Sexual selection and mate choice. In *Behavioural ecology* (eds J. R. Krebs and N. B. Davies), pp. 179–202. Oxford University Press, Oxford.

Ryan, M. J. (1998). Sexual selection, receiver biases, and the evolution of sex differences. *Science*, 281, 1999–2003.

Sakaluk, S. K. (2000). Sensory exploitation as an evolutionary origin to nuptial food gifts in insects. *Proceedings of the Royal Society of London B*, 267, 339–43.

Savage-Rumbaugh, S., Shanker, S., and Taylor, T. (1998). *Apes, language and the human mind*. Oxford University Press, Oxford.

Scheib, J. E., Gangstead, S. W., and Thornhill, R. (1999). Facial attractiveness, symmetry and cues of good genes. *Proceedings of the Royal Society of London B*, 268, 1913–17.

Schiestl, F. P., Ayasse, M., Paulus, H. F., Lofsted, C., Hansson, B. S., Ibarra, F., and Francke, W. (1999). Orchid pollination by sexual swindle. *Nature*, 399, 421–2.

Sebeok, T. A. (1977). *How animals communicate*. Indiana University Press, Bloomington.

Seeley, T. D. (1998). Thoughts on information and integration in honey bee colonies. *Apidologie*, 29, 67–80.

Semple, S. and McComb, K. (1996). Behavioural deception. *Trends in Ecology and Evolution*, 11, 434–7.

Senar, J. C., Camerino, M., Copete, J. L., and Metcalfe, N. B. (1993). Variation in black bib of the Eurasian Siskin (*Carduelis spinus*) and its role as a reliable badge of dominance. *Auk*, 110, 924–7.

Senar, J. C. and Camerino, M. (1998). Status signalling and the ability to recognize dominants: an experiment with Siskins (*Carduelis spinus*). *Proceedings of the Royal Society of London B*, 265, 1515–20.

Seyfarth, R. M. and Cheney, D. L. (1980). The ontogeny of Vervet Monkey alarm-calling behavior: a preliminary report. *Zeitschrift fur Tierpsychologie*, 54, 37–56.

Seyfarth, R. M. and Cheney, D. L. (1986). Vocal development in Vervet Monkeys. *Animal Behaviour*, 1640–58.

Seyfarth, R. M., Cheney, D. L., and Marler, P. (1980). Monkey responses to three different alarm calls: evidence for predator classification and semantic communication. *Science*, 210, 801–3.

Shaw, K. L. (1995). Phylogenetic tests of the sensory exploitation model of sexual selection. *Trends in Ecology and Evolution*, 10, 117–20.

Sherman, P. W. (1977). Nepotism and the evolution of alarm calls. *Science*, 197, 1246–53.

Sherman, P. W. (1981). Kinship, demography, and Belding's Ground Squirrel nepotism. *Behvioural Ecology and Soiobiology*, 8, 251–9.

Sherman, P. W., Reeve, H. K., and Pfennig, D. W. (1997). Recognition systems. In *Behavioural ecology: an evolutionary approach* (eds J. R. Krebs and N. B. Davies), 4th edn. Blackwell, Oxford.

Sheppard, P. M. (1958). *Natural selection and heredity*. Hutchinson, London.

Shields, W. M. (1977). The social significance of avian winter plumage variability: a comment. *Evolution*, 31, 905–7.

Shreeve, T. G. (1987). The mate location behaviour of the male Speckled Wood Butterfly, *Pararge aegeria*, and the effect of phenotypic differences in hind-wingspotting. *Animal Behaviour*, 35, 682–90.

Silk, J. B. (1999). Why are infants so attractive to others? The form and function of infant handling in Bonnet Macaques. *Animal Behaviour*, 57, 1021–32.

Silk, J. B. (2001). Grunts, girneys and good intentions: the origins of commitment in nonhuman primates. In *Evolution and capacity for commitment* (ed. R. Nesse), pp. 138–57. Russell Sage Press, New York.

Silk, J. B., Kaldor, E., and Boyd, R. (2000). Cheap talk when interests conflict. *Animal Behaviour*, 59, 423–32.

Silverman, H. B. and Dunbar, M. J. (1980). Aggressive tusk use by the Narwhal *Monodon monoceros L. Nature*, 284, 57–8.

Sinervo, B. and Lively, C. M. (1996). The rock-scissors-paper game and the evolution of alterntive male strategies. *Nature*, 380, 240–3.

Singer, F., Riechert, S. E., Xu, H. F., Morris, A. W., Becker, E. Hale, J. A., and Noureddine, M. M. (2000). Analysis of courtship success in the funnel-web spider *Agelenopsis aperta*. *Behaviour*, 137, 93–17.

Slater, P. J. B. (1983). The study of communication. In *Animal behaviour* (eds P. G. Bateson and P. J. B. Slater) vol. 2. Blackwell, Oxford.

Slater, P. J. B. and Ince, S. A. (1982). Song development in Chaffinches: what is learnt and when? *Ibis*, 124, 21–6.

Smith, V. A., King, A. P., and West, M. J. (2000). A role of her own: female cowbirds, *Moluthrus ater*, influence the development and outcome of song learning. *Animal Behaviour*, 60, 599–609.

Sneddon, L. U., Huntingford, F. A., and Taylor, A. C. (1997). Weapon size versus body size as a predictor of winning fights between Shore Crabs, *Carcinus maenas* (L). *Behavioural Ecology and Sociobiology*, 41, 237–42.

Soha, J. A. and Marler, P. (2000). A species-specific acoustic cue for selective song learning in the White-crowned Sparrow. *Animal Behaviour*, 60, 297–306.

Solberg, E. J. and Ringsby, T. H. (1997). Does male badge size signal status in small island polulations of House Sparrows, *Passer domesticus*? *Ethology*, 103, 177–86.

Stamps, J. A. and Krishnan, V. V. (1998). Territory acquisition in lizards. IV. Obtaining high status and exclusive home ranges. *Animal Behaviour*, 55, 461–72.

Stamps, J. A. and Krishnan, V. V. (1999). A learning-based model of territory establishment. *Quarterly Review of Biology*, 74, 291–318.

Steele, R. H. (1986*a*). Courtship feeding in *Drosophila subobscura*. I. The nutritional significance of courtship feeding. *Animal Behaviour*, 34, 1087–98.

Steele, R. H. (1986*b*). Courtship feeding in *Drosophila subobscura*. II. Courtship feeding by males influences female mate choice. *Animal Behaviour*, 34, 1099–108.

Struhsaker, T. T. (1967). Auditory communication among Vervet Monkeys (*Cercopithecus aethiops*). In *Social communication among primates* (ed. S. A. Altmann). University of Chicago Press, Chicago.

Struhsaker, T. T. (1975). *The Red Colobus monkey*. University of Chicago Press, Chicago.

Sullivan, K. (1985). Selective alarm-calling by Downy Woodpeckers in mixed-species flocks. *Auk*, 102, 184–87.

Swaddle, J. E. and Cuthill, I. C. (1995). Asymmetry and human facial attractivenes—symmetry may not always be beautiful. *Proceedings of the Royal Society of London B*, 261, 111–16.

Swaddle, J. P. and Witter, M. S. (1995). Chest plumage, dominance and fluctuating asymmetry in female starlings. *Proceedings of the Royal Society of London B*, 260, 219–23.

Takahata, Y. (1990). Social relationships among adult males. In *The Chimpanzees of the Mahale Mountains* (ed. T. Nishida). University of Tokyo Press, Tokyo.

Tarsitano, M., Jackson, R. R., and Kirchner, W. H. (2000). Signals and signal choices made by the araneophagic jumping spider *Portia fimbriata* while hunting the orb-weaving web-spiders *Zygiella x-notata* and *Zosis geniculatus*. *Ethology*, 106, 595–615.

Taylor, P. W., Hasson, O., and Clark, D. L. (2000). Body postures and patterns as amplifiers of physical condition. *Proceedings of the Royal Society of London B*, 267, 917–22.

Thapar, V. (1986). *Tigers: portrait of a predator*. Collins, London.

Thomas, R. J. (2002). The costs of singing in nightingales. *Animal Behaviour*, 63, 959–66.

Thornhill, R. (1976). Sexual selection and nuptial feeding behaviour in *Bittacus apicalis* (Insecta: Mecoptera). *American Naturalist*, 110, 529–48.

Thorpe, W. H. (1958). The learning of song patterns by birds, with special reference to the Chaffinch *Fringilla coelebs*. *Ibis*, 100, 535–70.

Tibbets, E. A. (2002). Visual signals of individual identity in the wasp *Polistes fuscatus*. *Proceedings of the Royal Society of London B*, 269, 1423–8.

Tinbergen, N. (1948). Social releasers and the experimental method required for their study. *Wilson Bulletin*, 60, 6–51.

Tinbergen, N. (1951). *The study of instinct*. Clarendon Press, Oxford.

Tinbergen, N. (1952). Derived activities: their causation, biological significance, origin and emancipation during evolution. *Quarterly Review of Biology*, 27, 1–32.

Tinbergen, N. (1953). *The Herring Gull's world*. Collins, London.

Tobias, J. (1997). Asymmetric territorial contests in the European Robin: the role of settlement costs. *Animal Behaviour*, 54, 9–21.

Tomasello, M. and Call, J. (1997). *Primate cognition*. Oxford University Press, New York.

Tomkins, J. L. and Simmons, L. W. (1999). Heritability of size but not symmetry in a sexually selected trait chosen by female earwigs. *Heredity*, 82, 151–7.

Tramer, E. J. (1994). Feeder access: deceptive use of alarm calls by a White-breasted Nuthatch. *Wilson Bulletin*, 106, 573.

Trivers, R. L. (1971). The evolution of reciprocal altruism. *Quarterly Review of Biology*, 46, 35–57.

Trivers, R. L. (1974). Parent–offspring conflict. *American Zoologist*, 14, 249–64.

Turner, J. R. G. (1984). Darwin's coffin and Dr Pangloss—do adaptationist models explain mimicry? In *Evolutionary ecology* (ed. B. Shorrocks), pp. 313–61. Blackwell, Oxford.

Vahed, K. (1998). The function of nuptial feeding in insects—review of empirical studies. *Biological Reviews*, 73, 43–78.

van Lawick-Goodall, H. and van Lawick-Goodall, J. (1970). *Innocent killers*. Collins, London.

van Someren, V. G. L. (1956). *Days with birds: studies of habits of some East African birds*. Chicago University Press, Chicago.

van Tets, G. F. (1965). A comparative study of some social communication patterns in the Pelecaniformes. *American Ornithologists Union Ornithological Monographs*, 2, 1–88.

Veiga, J. P. (1993). Badge size, phenotypic quality and reproductive success in the House Sparrow—a study in honest advertisement. *Evolution*, 47, 1161–70.

Versluis, M., Schmitz, B. von der Heydt, A., and Lohse, D. (2000). How Snapping Shrimp snap: through cavitating bubbles. *Science*, 289, 2114–17.

Viljugrein, H. (1997). The cost of dishonesty. *Proceedings of the Royal Society of London B*, 264, 815–21.

Vincent, A., Ahnesjo, J., and Berglund, A. (1994). Operational sex ratios and behavioural sex-differences in a pipefish population. *Behavioural Ecology*, 34, 435–42.

von Frisch, K. (1950). *Bees: their vision, chemical senses and language*. Cornell University Press, New York.

Waddington, C. H. (1957). *The strategy of the genes*. Macmillan, New York.

Walther, F. R. (1969). Flight behaviour and avoidance of predators in the Thomson's Gazelle (*Gazella thomsoni* Guenther 1884). *Behaviour*, 34, 184–221.

Watt, D. J. (1986). Relationship between plumage variability, size and sex to social-dominance in Harris' Sparrows. *Animal Behaviour*, 34, 16–27.

Weary, D. M., Guilford, T. C., and Weisman, R. G. (1993). A product of discriminative learning may lead to female preferences for elaborate males. *Evolution*, 47, 333–36.

West, M. J. and King, A. P. (1988). Female visual-displays affect the development of male song in the Cowbird. *Nature*, 334, 244–6.

West-Eberhard, M. J. (1979). Sexual selection, competition and evolution. *Proceedings of the American Philosophical Society*, 123, 222–34.

Whitekiller, R. R., Westneat, D. F., Schwagmeyer, P. L., and Mock, D. W. (2000). Badge size and extra-pair fertilizations in the House Sparrow. *Condor*, 102, 342–8.

Whiten, A. (1997). The Machiavellian mindreader. In *Machiavelian intelligence II* (eds A. Whiten and R. W. Byrne), pp. 144–73. Cambridge University Press, Cambridge.

Whitfield, D. P. (1986). Plumage variability and territoriality in breeding Turnstones *Arenaria interpres*: status signalling or individual recognition. *Animal Behaviour*, 34, 1471–82.

Whitfield, D. P. (1987). Plumage variability, status signalling, and individual recognition in avian flocks. *Trends in Ecology and Evolution*, 2, 13–18.

Whitlock, M. C. and Fowler, K. (1997). The instability of studies of instability. *Journal of Evolutionary Biology*, 10, 63–7.

Wiley, R. H. (1983). The evolution of communication: information and manipulation. In *Communication* (eds T. R. Halliday and P. J. B. Slater), pp. 82–113. Blackwell, Oxford.

Wilkinson, G. S. and Reillo, P. R. (1994). Female choice response to artificial selection on an exaggerated male trait in a stalk-eyed fly. *Proceedings of the Royal Society of London B*, 255, 1–6.

Wilson, J. (1994). Variation in initiator strategy in fighting by silvereyes. *Animal Behaviour*, 47, 153–62.

Wilson, J. D. (1992*a*). Correlates of agonistic display by Great Tits *Parus major*. *Behaviour*, 121, 168–214.

Wilson, J. D. (1992*b*). A reassessment of the significance of status signalling in populations of wild Great Tits, *Parus major*. *Animal Behaviour*, 43, 999–1009.

Wimmer, H. and Perner, J. (1983). Beliefs about belief: representation and constraining function of wrong beliefs in young children's understanding of deception. *Cognition*, 13, 103–28.

Winter, P., Handley, P., Ploog, D., and Schott, D. (1973). Ontogeny of Squirrel Monkey calls under normal conditions and under acoustic isolation. *Behaviour*, 47, 230–9.

Wrangham, R. (1986). Feeding behaviour of Chimpanzees in Gombe National Park, Tanzania. In *Primate ecology* (ed. T. H. Clutton-Brock), pp. 503–8. Academic Press, London.

Wright, J. (1997). Helping at the nest in Arabian Babblers: signalling social status or sensible investment in chicks? *Animal Behaviour*, 54, 1439–48.

Wright, J. (1999). Altruism as a signal—Zahavi's alternative to kin selection and reciprocity. *Journal of Avian Biology*, 30, 108–15.

Zahavi, A. (1971). The social behaviour of the White Wagtail *Motacilla alba alba* wintering in Israel. *Ibis*, 113, 203–11.

Zahavi, A. (1975). Mate selection—a selection for a handicap. *Journal of Theoretical Biology*, 53, 205–14.

Zahavi, A. (1977). Reliability in communication systems and the evolution of altruism. In *Evolutionary ecology* (eds. B. Stonehouse and C. M. Perrins), pp. 253–9. Macmillan, London.

Zahavi, A. (1990). Arabian Babblers: the quest for social status in a cooperative breeder. In *Cooperative breeding in birds: long-term studies of ecology and behaviour* (eds P. Stacey and W. Koenig), pp. 105–30. Cambridge, Cambridge University Press.

Zahavi, A. (1995). Altruism as a handicap—limitations of kin selection and reciprocity. *Journal of Avian Biology*, 26, 1–3.

Zahavi, A. and Zahavi, A. (1997). *The handicap principle: a missing piece of Darwin's puzzle*. Oxford University Press, New York.

Zuberbuhler, K. (2000). Interspecies semantic communication in two forest primates. *Proceedings of the Royal Society of London B*, 267, 713–18.

Index

ability to
communicate 115–16, 133
produce the correct signal 116
acoustic differences 115
'action-response'
games 11–12, 30, 91, 100, 122
model 20
Adams, E. S. 88
African Grey Parrots 113
African Hunting Dogs 127
African Swallowtail Butterfly 87
African Wild Dogs 62
aggressive signals 19, 80–1
Albon, S. D. 1
altruistic behaviour 22, 43
'altruistic punishment' 124–6
in humans 125–6
American Goldfinches 79, 101
Andersson, M. 18, 80
angler fish 5, 10
antelope 57, 61, 67
'anti-Bourgeois' strategy 55–6
aposematism 4, 6
Arabian Babblers 30
Arctiid moths 6
'assessment signals' 9
asymmetry 12–13, 30, 38–9, 55, 64–5, 94, 96, 98, 104, 110, 122, 125
Atlantic Gannet 70
Aubin, T. 73
Australian Magpies 58
avian example 92

baboon alarm 114
baboons 120, 124
Backwell, P. R. Y. 88
'badges of status' 6, 19, 90, 92–8, 110
Bald Eagle 17, 48
Barber, I. 53
basilar papilla 82–3
Basolo, A. L. 81, 83
Bearded Tits 127
behavioural cues 69
intention movements 69–70
protective movements 70–1
displacement behaviour 69
Belding's Ground Squirrel 43
Bergland, A. 95
Bergstrom, C. T. 25, 27, 29, 31, 36
Bickerton, D. 133–4, 136
Bighorn Sheep 67
Black Stork 129
Bloom, P. 133, 136
Blue-footed Booby 70
Bluehead Wrasse 39
Bolas spiders 10
Bourgeois strategy 38, 54–6, 124
bowerbird 17
Bradbury, J. W. 1
Brakefield, P. M. 65
breast-to-breast display 8, 19
Breuker, C. F. 65
Brown, J. 101

Caldwell, R. L. 88
Canary chicks 36
Caro, T. M. 61
Caryl, P. 102
Cercopithecoid monkeys 114
Chacma Baboons 123–4
grunts as signals of friendliness 124
Cheetahs 62
Cheney, D. L. 113, 115, 118
chick begging 21, 27, 35–7
chicken 120
Chimpanzee 113, 125, 128–31
cultural inheritance 131

Chirrup calls 39
choice-based signals 14
Chomsky, N. T. 132
  Chiffchaff 11, 137
cichlid fish 51, 53, 59, 81, 91, 100, 102, 110
classification of 'intentionality'
  first-order 119
  second-order 119
  zero-order 119
Clutton-Brock, T. H. 1, 99
Collared Flycatchers 34, 93
Common Goldeneyes 72
'common interest' 32, 37, 39
Common Redpolls 101
Common Toads 1, 18, 50
  croak in 50
concept of a Nash Equilibrium 124
condition-dependent
  mating strategies 65
  expenditure 60
'consensus games' 90
continuously varying signallers
  model 22–5
conventional signals of need
  model of 97–8
'coordination games' 30, 37, 39
'cost' 7, 15–16, 21
  efficacy cost 7, 15–17, 27, 30, 41
  strategic cost 7–8, 15–17, 27, 30–1, 33
cost-free
  equilibrium 23
  models of 'intent' 122
  'pooling equilibria' 28
  signalling 20, 22–3, 25, 27–8, 30–2, 37, 98
  signals of intent 123
  theories 37
costly signals 19, 26, 28, 31–2, 36, 108, 111
courtship 88
  dances 51–2, 60
  feeding 4, 60
Coutlee, E. 101
Cowbirds 117
crabs 66–7, 88
Crested Auklet 93
Crested Tits 40
Crowned Eagle 114, 116
cues 3–6, 10, 15, 29, 41–3, 46–7, 54, 59, 64–5, 67–71, 78, 84, 86, 115
Cumming, J. M. 85
Curlew 11
Cuthill, I. C. 65

Dance Flies 85, 88
Darwin, C. 72
David, P. 34
Davies, N. B. 1, 59
Dawkins, M. S. 39, 41, 71, 73–4
Dawkins, R. 3, 10, 55, 74
deception 86–8
Dennett, D. C. 119
Dilger, W. 101
direct reputation 122
  evidence for 123–4
discrete signals 97–9, 110
discrete model 21–2
dishonest signals 37
displays of weapons 59, 66–7
dolphins 113, 135
domestic chicks 74
Double-crested Cormorants 127
Downy Woodpeckers 120
Dyed birds 99

Eco, U. 6
edible mimic 87
electromagenetic receptors 5
Elephant Seals 40
Emperor Penguin 73, 138
Empid flies 85
Endler, J. A. 73, 85
Enquist, M. 1, 7, 18, 50, 76, 81, 98, 101–4
Enquist's first model 19, 77
escalated fight 17, 19, 34–5, 37, 50, 56, 89–91, 93–4, 96, 101, 105, 109, 122
ESS models of badges 74, 94–5
Eurasian Nuthatches
  'conflict call' of 88
Eurasian Siskins 92
European Robin 55, 71, 138
European Starlings 93
eusocial hymenoptera 42
eusocial insect colonies 27
evolution of
  indices 47–8
  language 133–6
  signal form 68–89
  signals 9
evolutionary equilibrium 8, 11, 20, 37

Fallow Deer 47, 138
Fanshawe, J. H. 61–3
fiddler crabs *see* crabs

fish 5, 8, 10, 12, 19, 51, 69, 73–4, 81, 83–4, 86, 91–2, 100, 102–4, 110, 136
Fisher, R. A. 13, 59, 83, 85
Fisherian process 14
Fitch, W. T. 46
FitzGibbon, C. D. 61–3
fixed intensity signal 28
fluctuating asymmetry (FA) 4, 45, 63–6
formal model 3, 17–18, 60, 81, 91, 108
forms of signals 9–11
  mimicry 5–6, 10, 57–8, 86–7, 89
  sensory manipulation 10–12, 81–5
fruit flies 39
funnel-web spiders 4, 45–6, 52, 110
  vibrating a web 50

Gaint Deer 49
Galef, B. G. 6
game-theoretic models 17
Gazelles 61
Geist, V. 67
genetic analysis 106, 136
'genetic correlation' 83
genetic model 17–19
  of sexual selection 18
genetic variance 14, 17, 34, 53
Gintis, H. 100, 126
Giraldeau, L. A. 6
Gladiator Frogs 66
Golden-Mantled Ground Squirrel 43, 138
Gombe chimps 129, 131
'Good genes'
  models 14, 53
  theories of sexual selection 83
Goodall, J. 129
Gopnik, M. 133
Gosling, L. M. 57
Gould, S. J. 50
Grafen, A. 7–8, 18, 20, 24, 26, 56
Grafen's model 26
Great Crested Grebes 126
Greater Flamingo 128
Great Tits 6, 56, 87, 119
  belly stripe of 93
group displays 128
  interpretation 128–30
group-living primates 120
Guilford, T. 39, 41, 73–4

Haka phenomenon 127
Halliday, T. R. 1
Ham, A. D. 93
Hamilton, W. D. 14, 18, 22, 53–4
Hamilton's rule ($rb>c$) 43
Hamilton–Zuk
  hypothesis 53
  process 18
handicaps 1, 3, 8–9, 14–15, 17, 45, 54, 60, 65
  equilibrium 34
  history of 17
  model 27–8, 36–7
  theories 37
Harper, D. G. C. 1, 93–5
Harris's Sparrow 93, 99, 138
Hartley, I. R. 94–5
Hasson, O. 1, 5, 3, 47, 61
Hauser, M. D. 1, 5–6, 100, 117
Hawaiian honeycreeper 55
Hawk–Dove game 38
Hen Harrier 52, 138
Hennessey, D. 114
Honey Badgers 41
Honey Bees 41, 115
  dance of 11
honeyguides 41
Horned Grebes 127
hostile actor 123
House Finches 34, 92
House Sparrows 6, 39, 93, 95, 99
  bibs 94
human language 37, 113, 115, 130–6
  peculiarities of 132
Hurd, P. L. 101–2, 104
Huxley, J. S. 126–7

icon 10–11, 15, 69
'index' 3, 9, 11, 14–15, 30, 32, 60
  condition 48–50
  ownership 54–9
  performance 51–3
  quality 9, 45–67
  reliable 46
  RHP 61, 81
  size 50–1
'Irish Elk' 49
Isack, H. A. 41

jackals 114, 127
Jackendoff, R. 134, 136
Javanese Munias 80
Jennions, M. D. 79
Johnstone, R. A. 1, 26, 95

Killdeer 11, 138
Killer Whales 40
Kilner, R. M. 36
kin recognition
  altruistic behaviour 42–4
  mutual recognition 42
  nest-mate recognition 42
Krebs, J. R. 3, 10, 55–6, 70, 74
Krishnan, V. V. 109

Lachmann, M. 25, 27, 29, 31, 36–7
LaPorte, J. 6
Leimar, O. 103
lekking species 13
Leopards 114, 134
Lesser Flamingo 127–8, 138
Lesser Spotted Woodpeckers 130
Lingle, S. 61
linguistic competence 133, 136
lions 51, 114
lizards 72, 92, 109
Long-tailed Grass-finch 79–80
Lorenz, K. 3, 71, 129
Lyrebird 59, 138

'Machiavellian Intelligence' 120
Mackintosh, N. J. 77
Mahale chimps 131
Mallard 71, 138
mammalian sounds 45–7
Mandarin Duck 71
Marler, P. 74, 100
Marsh Tits 87
Marsh Warblers 58
Martial Eagle 114, 138
Mateo, J. M. 43
Matsuoka, S. 87
Maynard Smith, J. 1, 9, 18–9, 35, 38, 67, 93–5, 105–6
McComb, K. E. 46, 51
McKay, H. V. 57
Meyer, A. 83
Michael Cullen 70
mimicry 5–6, 10, 57–8, 86–9
  'Batesian' 4–5, 10, 87
  and cheating 86–9
  vocal 57–8
Møller, A. P. 63–4, 87, 119
monkeys 100
Moremi reserve in Botswana 124
Morris, D. J. 68, 101
'motivational analysis' 101
'motivational' model 100–1, 105–8
mouth-wrestling of cichlid fishes 53
Mule Deer 61
Mullerian mimicry 10
Musk Ducks 71–2
musth in elephants 34–5
'mutants' 28
  'bluffing' 95, 98
  'modest' 95–6
mutual displays 125–30

Naked Mole Rats 99
Narwhals 66
'negotiation game' 91, 101, 108
nightingale 16, 139
  song of 32
Noldeke, G. 24
non-equilibrium dynamics 76, 87
non-signalling equilibrium 29, 31
non-symmetrical characteristics 65
Norris, K. 95
Northern Fulmars 7, 19
'null model' 98
nuptial gifts in insects 52, 84–5

olfactory signals 57
optimization approach 11
orangutans 72, 130
orb-web spiders 10, 101
Osorio, D. 93
Owens, I. P. F. 94–5
Owings, D. 114
oystercatchers 11

Painted Redstarts 87
'pant-hoot' call 129
paper wasps *see* wasps
parasites 18, 53–4, 59, 93, 117
Parker, G. A. 9, 99
passerine birds 13, 75, 101
Paton, D. 102
peacock 7, 32, 59, 139
  tail of 17
'peak shift' 77, 81
Peewit 11
perceptual error 25–6
Peregrine Falcon 52, 139
performance-based signals 1–2, 11, 14
performance index
  courtship dance 52

courtship feeding 52
stotting 52
performance indicators 60
Perner, J. 120
Perrett, D. I. 65
Philip Sidney game 17, 20–7, 29–30
Philip Sidney, Sir 20
'phonemes' 113, 134
Pied Wagtail 58, 139
Pinker, S. 133, 136
pipefishes 12
Plenge, M. 80
Pomiankowski, A. 8, 18–19
Poole, J. H. 34
'pooling equilibria' 27–8, 36–7
Popp, J. 101
Price, G. R. 38, 67
problem of reliability 6–9
Proctor, H. C. 84
protolanguage 133–6
protracted contests 12, 50, 100–3, 105–7, 110
punishment 99–100
of false signals 30

qualitatively novel signals 78

rain dance 129
Reby, D. 46
receivers 3, 10, 22, 29, 31, 40, 46, 68, 7–14, 81–3, 86–7, 102, 119
as 'mind-readers' 3, 74
Red Colobus Monkeys 114
Red Deer 1
roaring of 45–6, 50
stags 18, 47, 49, 67
Red Howler Monkey 46
Red-winged Blackbirds 93
relatedness 22–7, 29, 41–3, 115, 120
reliability of animal signal 32
reliable indicator
dancing 61
repeated interactions 30, 38–9, 44, 100
resource-holding potential RHP 50–1, 61, 67, 95, 103, 110
response to novel signals 77–80
'revealing handicap' 17–18
Reyer, H.-U. 41
Rhesus Macaques 46, 115, 123
Rhesus monkeys 115
Rice, W. 77
Riechert, S. E. 4, 50, 105–6
ritualization 15, 68–73
of movements 10
pupil dilation 68
'Rock-Scissors-Paper' game 97
Rodd, H. F. 86
Roe Deer 57
Rohwer, F. C. 99
Rohwer, S. 99
Ruddy Turnstones 92
Rufous-sided Towhees 73
'runaway process' 13
Ryan, M. J. 81, 83

Sage Grouse 54, 139
Sakaluk, S. K. 85
salticid spider 10, 49
Samuelson, L. 24
Scarlet-tufted Malachite Sunbirds 93
scent-marking 57
scorpionfly 84
sea horses 12
Sebeok, T. A. 80
Seeley, T. D. 41, 115
'sensory bias' 74, 82, 95
hypothesis 83
non-adaptive 13
sensory exploitation 14–15, 35, 85
sensory manipulation 12, 81, 85
examples 85–6
sequential assessment
game 102
model 103, 107
sex-role reversal 12
sexual selection 12–15, 19
'condition-dependence' 34
courtship 12
models of 3
paternal care 13
'runaway' theory of 83
sexually attractive sons 13
Seyfarth, R. M. 113, 115, 118
Shore Crabs 67 *see also* crabs
sibling competition 36
signals 15, 18
choice 18
and cues 3–6
during contests 59–60, 90–111
efficacy of 73–4
honesty of 121–4, 29, 37
of intent 121
in mate choice 59–60
minimal-cost 15, 32–44, 90, 123
non-variability of 28

nt.)
rship 57–8
rformance 1, 11, 14, 18
'persuit deterrance' 61
in primates 112–36
reliable 2, 31, 56, 60, 96, 123–4
'risky' 8
ritualized 71
role of 112
in social animals 112–36
social reputation of 121–4
strategic 32–44
signallers as 'manipulators' 3, 74
signalling equilibrium 9, 18–20, 22–4, 31
signalling systems 8–9, 37, 80, 89, 91
Silk, J. B. 38, 122–3
Silver-backed Jack 127, 139
Silver-washed Fritillary butterfly 52
snakes 114
Snapping Shrimps 50, 67
Sneddon, L. U. 67
'source-filter' theory 45–6
sparrowhawks 40
spider fights 105–8
Squirrel Monkeys 116
Stag Beetles 66
stalk-eyed flies 7, 14, 33, 45, 64
eye-stalks of 59
Stamps, J. A. 109
Steele, R. H. 84
Stellar's Jay 101
stotting 61–3
'strategic costs' 7–8, 17, 30–1
of signalling 8
Struhsaker, T. T. 114
'supernormal stimulus' 11, 77
Swaddle, J. E. 65
swallows 63–5, 127
symbol 11, 15, 59, 69, 134–5

Tai chimps 131
Takahata, Y. 129
tape-tutoring experiments 117
Taylor, P. W. 48
territorial behaviour 108
territorial mammals 57
Thapar, V. 1
theoretical models 26, 110
theory of
costly signalling 16–31
mind 120
sensory manipulation 11
Thornhill, R. 63, 84
Thorpe, W. H. 117
Three-spined Stickleback 53, 139
Tibbetts, E. A. 43
tigers 1, 46–7, 50
Tinbergen, N. 9, 77
Tobias, J. 55
Trivers, R. L. 35
two-voice system 73
Tmúngara Frog 82

unfakeable index 67
unfakeable signals 1, 11, 50, 110
'unfamiliar human' alarm 114
Unicorn 66

Vahed, K. 84
varied signals 100–2
Vehrencamp, S. L. 1
Vervet Monkeys 112–3, 115, 118
alarm calls 121
'Leopard' alarm of 134

waggle dance 41
'War of Attrition' 96–8
game 104
'War of the Sexes' game 37
wasps 41–2, 66, 99
'weed dance' 127
West-Eberhard, M. J. 18
Western Grebes 127
White-browed Robin 58, 140
White Stork 129, 140
White Wagtails 58, 140
Whiten, A. 121
Wild Dogs 62
Wiley, R. H. 73
Willow Tits 87
Wimmer, H. 120
Winter, P. 116
wolves 66
Wood Ducks 71
worm moths 10
Wright, J. 30

Zahavi, A. 1, 6, 16–18, 24, 29–30, 48, 54, 72
Zahavi's handicap principle 17
Zahavi's verbal model 3
Zebra Finches 16, 79
Zuberbuhler, K. 114
Zuk, M. 14, 18, 53–4